Macmillan/McGraw-Hill • Glencoe

Grade 3

Math Triumphs

Book 3: Geometry

Authors

Basich Whitney • Brown • Dawson • Gonsalves • Silbey • Vielhaber

Macmillan/McGraw-Hill
Glencoe

Photo Credits

The McGraw·Hill Companies

Macmillan/McGraw-Hill
Glencoe

Copyright © 2009 by The McGraw-Hill Companies, Inc. All rights reserved. No part of this publication may be reproduced or distributed in any form or by any means, or stored in a database or retrieval system, without the prior written consent of The McGraw-Hill Companies, Inc., including, but not limited to, network storage or transmission, or broadcast for distance learning.

Send all inquiries to:
Macmillan/McGraw-Hill • Glencoe/McGraw-Hill
8787 Orion Place
Columbus, OH 43240-4027

ISBN: 978-0-07-888200-5
MHID: 0-07-888200-1

Printed in the United States of America.

2 3 4 5 6 7 8 9 10   066   16 15 14 13 12 11 10 09 08

Math Triumphs
Grade 3, Book 3

Book 1

Book 2

Book 3

Authors and Consultants

CONSULTING AUTHORS

Frances Basich Whitney
Project Director, Mathematics K-12
Santa Cruz County Office of Education
Capitola, California

Kathleen M. Brown
Math Curriculum Staff Developer
Washington Middle School
Long Beach, California

Dixie Dawson
Math Curriculum Leader
Long Beach Unified
Long Beach, California

Philip Gonsalves
Mathematics Coordinator
Alameda County Office of Education
Hayward, California

Robyn Silbey
Math Specialist
Montgomery County Public Schools
Gaithersburg, Maryland

Kathy Vielhaber
Mathematics Consultant
St. Louis, Missouri

CONTRIBUTING AUTHORS

Viken Hovsepian
Professor of Mathematics
Rio Hondo College
Whittier, California

FOLDABLES Study Organizer **Dinah Zike**
Educational Consultant
Dinah-Might Activities, Inc.
San Antonio, Texas

CONSULTANTS

Assessment

Donna M. Kopenski, Ed.D.
Math Coordinator K–5
City Heights Educational Collaborative
San Diego, California

Instructional Planning and Support

Beatrice Luchin
Mathematics Consultant
League City, Texas

ELL Support and Vocabulary

ReLeah Cossett Lent
Author/Educational Consultant
Alford, Florida

Reviewers

Each person reviewed at least two chapters of the Student Study Guide, providing feedback and suggestions for improving the effectiveness of the mathematics instruction.

Dana M. Addis
Teacher Leader
Dearborn Public Schools
Dearborn, MI

Renee M. Blanchard
Elementary Math Facilitator
Erie School District
Erie, PA

Jeanette Collins Cantrell
5th and 6th Grade Math Teacher
W.R. Castle Memorial Elementary
Wittensville, KY

Helen L. Cheek
K-5 Math Specialist
Durham Public Schools
Durham, NC

Mercy Cosper
1st Grade Teacher
Pershing Park Elementary
Killeen, TX

Bonnie H. Ennis
Math Coordinator
Wicomico County Public Schools
Salisbury, MD

Sheila A. Evans
Instructional Support Teacher - Math
Glenmount Elementary/Middle School
Baltimore, MD

Lisa B. Golub
Curriculum Resource Teacher
Millennia Elementary
Orlando, FL

Donna Hagan
Program Specialist - Special Programs
    Department
Weatherford ISD
Weatherford, TX

Russell Hinson
Teacher
Belleview Elementary
Rock Hill, SC

Tania Shepherd Holbrook
Teacher
Central Elementary School
Paintsville, KY

Stephanie J. Howard
3rd Grade Teacher
Preston Smith Elementary
Lubbock, TX

Rhonda T. Inskeep
Math Support Teacher
Stevens Forest Elementary School
Columbia, MD

Albert Gregory Knights
Teacher/4th Grade/Math Lead Teacher
Cornelius Elementary
Houston, TX

Barbara Langley
Math/Science Coach
Poinciana Elementary School
Kissimmee, FL

David Ennis McBroom
Math/Science Facilitator
John Motley Morehead Elementary
Charlotte, NC

Jan Mercer, MA; NBCT
K-5 Math Lab Facilitator
Meadow Woods Elementary
Orlando, FL

Rosalind R. Mohamed
Instructional Support Teacher - Math
Furley Elementary School
Baltimore, MD

Patricia Penafiel
Teacher
Phyllis Miller Elementary
Miami, FL

Lindsey R. Petlak
2nd Grade Instructor
Prairieview Elementary School
Hainesville, IL

Lana A. Prichard
District Math Resource Teacher K-8
Lawrence Co. School District
Louisa, KY

Stacy L. Riggle
3rd Grade Spanish Magnet Teacher
Phillips Elementary
Pittsburgh, PA

Wendy Scheleur
5th Grade Teacher
Piney Orchard Elementary
Odenton, MD

Stacey L. Shapiro
Teacher
Zilker Elementary
Austin, TX

Kim Wilkerson Smith
4th Grade Teacher
Casey Elementary School
Austin, TX

Wyolonda M. Smith, NBCT
4th Grade Teacher
Pilot Elementary School
Greensboro, NC

Kristen M. Stone
3rd Grade Teacher
Tanglewood Elementary
Lumberton, NC

Jamie M. Williams
Math Specialist
New York Mills Union Free School District
New York Mills, NY

Contents

CHAPTER

8

Measurement and Geometry

CHAPTER 9 Data Analysis

Home Connection

English

Dear Family,
Today our class started **Chapter 7, Geometry.** In this chapter, I will learn about the names and properties of two- and three-dimensional figures.

Love, _____

Spanish

Estimada familia:
Hoy en clase comenzamos el **Capítulo 7, Geometría.** En este capítulo aprenderé los nombres y propiedades de las figuras bidimensionales y tridimensionales.

Cariños, _____

Help at Home

You can help your child learn about two- and three-dimensional figures by asking him or her to name the shape of common household objects such as cubes of sugar, plates, or throw rugs. You can also help your child by describing the properties of these objects.

Ayude en casa

Usted puede ayudar a su hijo(a) a aprender acerca de las figuras bidimensionales y tridimensionales preguntándole el nombre de objetos de uso común en el hogar, como azúcar, platos o tapetes. También puede ayudar a su hijo(a) al describirle las propiedades de estos objetos.

Math Online Take the chapter Quick Check quiz at macmillanmh.com.

Name _____

Get Ready

Draw an X on the figure that is different.

1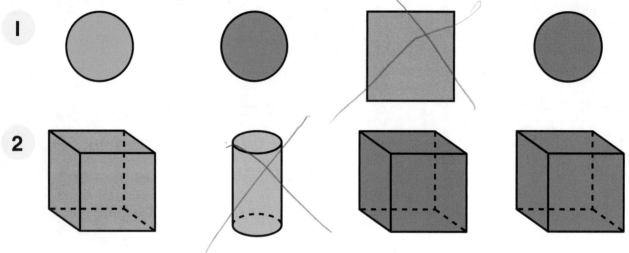

2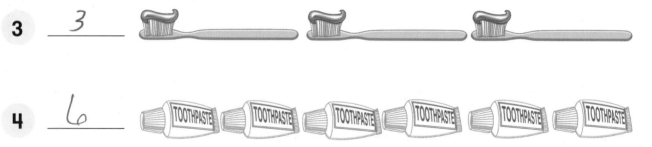

Count. Tell how many.

3 ___3___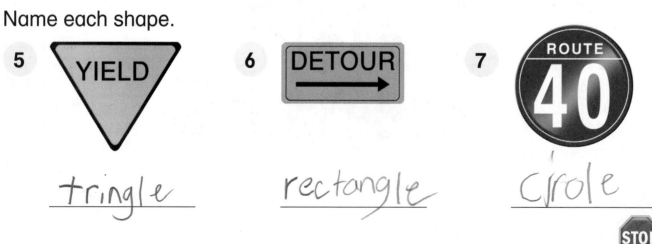

4 ___6___

Name each shape.

5 YIELD

tringle

6 DETOUR

rectangle

7 ROUTE 40

circle

STOP

Name _____

Three-Dimensional Figures

Key Concept

Each object is a three-dimensional figure.

cone

cube

cylinder

sphere

Vocabulary

cone   a three-dimensional figure that narrows to a point from a circular base

cube   a three-dimensional figure with 6 square faces

cylinder   a three-dimensional figure that looks like a can

sphere   a three-dimensional figure that has the shape of a round ball

Cubes and cylinders are easy to stack. Spheres, cones, and cylinders can roll.

Circle the object that is similar to the figure.

Step 1   The figure will stack and roll.

Step 2   The figure is a cylinder.

Step 3   The recycle bin will also stack and roll.

Answer   The recycle bin is similar to the cylinder.

Step-by-Step Practice

Circle the object that is similar to the figure.

Step 1   The figure will _____.

Step 2   The figure is a _____.

Step 3   The _____ will also _____.

Answer   The _____ is similar to the _____.

Name _____

▶ Guided Practice

Circle the object that is similar to each figure.

1.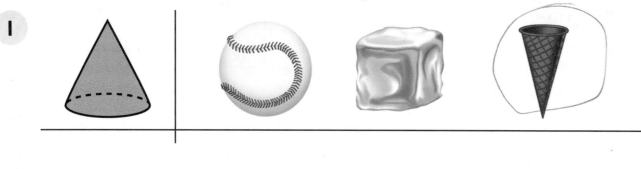

2.

Problem-Solving Practice

3. Liam brought a wrapped present to a party. What three-dimensional figure is the present?

   Understand Underline key words.

   Plan Use a model.

   Solve Look at models of three-dimensional figures.

   The present is a _____.

   Check Will the present roll or stack like the figure?

GO on

▶ Practice on Your Own

Name each three-dimensional figure.

4  _Sphere_

5 _Cylinder_

Circle the object that is similar to each figure.

6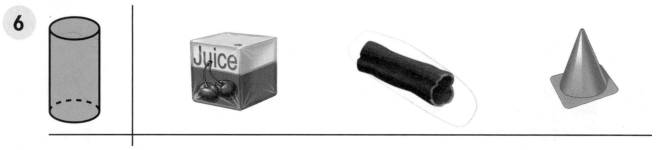

7

8 **WRITING IN ▶MATH** What three-dimensional figure is similar to the party hat? How do you know?

Cone △

Vocabulary Check Complete.

9 A three-dimensional figure that is similar to a soccer ball is called a _____.

Name _____

Faces and Edges

Key Concept

Look at the **rectangular prism** below.
There are 6 flat surfaces called **faces**.

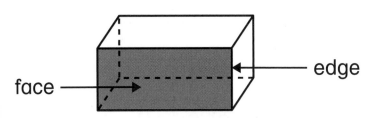

face ⟶   ⟵ edge

There are 12 lines where the faces touch.
These are called **edges**.

Vocabulary

rectangular prism   a three-dimensional
figure with faces that are rectangles

face   the flat part of a three-dimensional
figure; a square is a face of a cube

edge   the line where two sides or faces
meet

pyramid   a three-dimensional figure
with a polygon as a base and triangular-
shaped faces

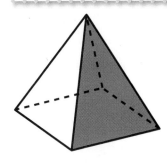

This pyramid has five faces.
Four faces are triangles.
One face is a square.

Example

Circle the figure that is similar to the blue face.

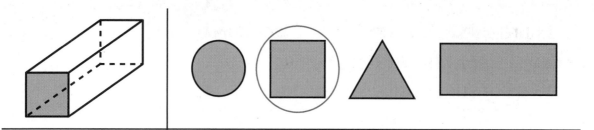

Step 1   The face has 4 edges.

Step 2   The second figure has the same shape.

Answer   I circled the second figure.

Step-by-Step Practice

Circle the figure that is similar to the blue face.

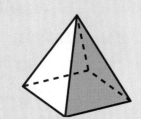 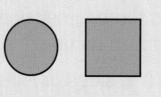

Step 1   The face has _____ edges.

Step 2   The _____ figure has the same shape.

Answer   I circled the _____ figure.

Name _____

▶ Guided Practice

Circle the figure that is similar to the blue face.

1

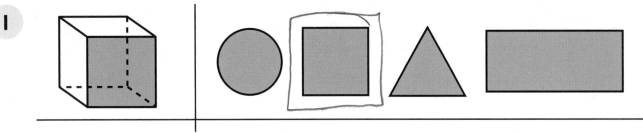

2

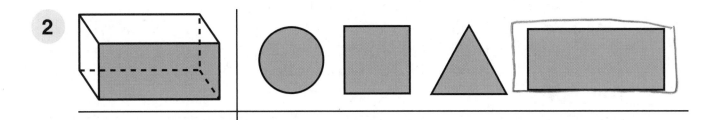

Problem-Solving Practice

3 I am a solid figure with 4 triangular faces that meet at a point. What figure am I?

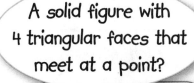

A solid figure with 4 triangular faces that meet at a point?

Understand   Underline key words.

Plan   Use a model.

Solve   Look at the three-dimensional figure models. Find the figure that matches the clues.

The figure is a _____.

Check   Count the triangular faces.

GO on

▶ Practice on Your Own

Circle the figure that is similar to the blue face.

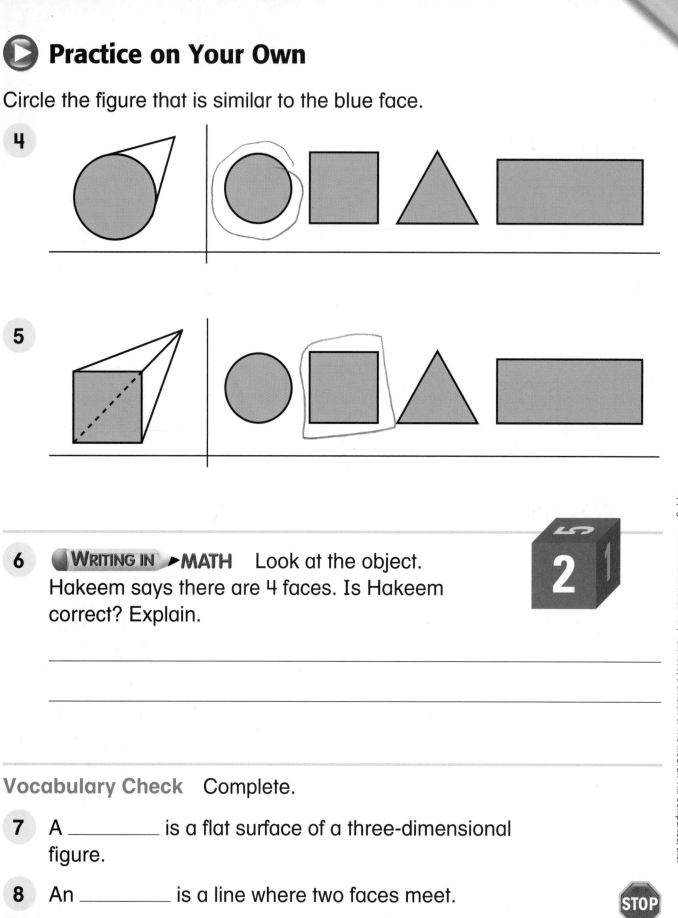

4

5

6 **WRITING IN ▶MATH**   Look at the object. Hakeem says there are 4 faces. Is Hakeem correct? Explain.

_____

_____

Vocabulary Check   Complete.

7 A _____ is a flat surface of a three-dimensional figure.

8 An _____ is a line where two faces meet.

Name _____

Progress Check 1 (Lessons 7-1 and 7-2)

1 Circle the object that is similar to the figure.

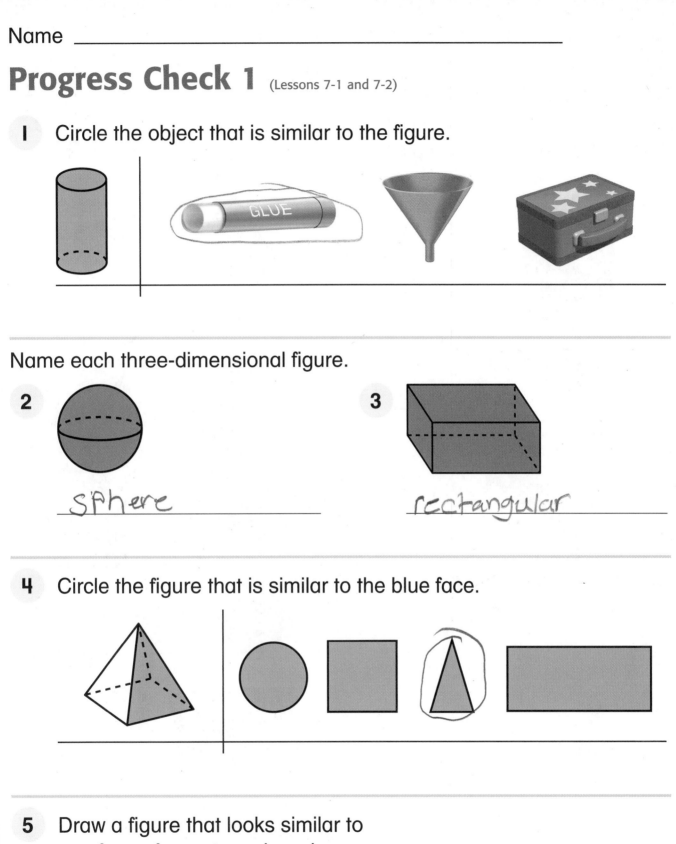

Name each three-dimensional figure.

2

sphere

3

rectangular

4 Circle the figure that is similar to the blue face.

5 Draw a figure that looks similar to one face of a rectangular prism.

Name _____

Draw a line to match each three-dimensional figure to an object or description.

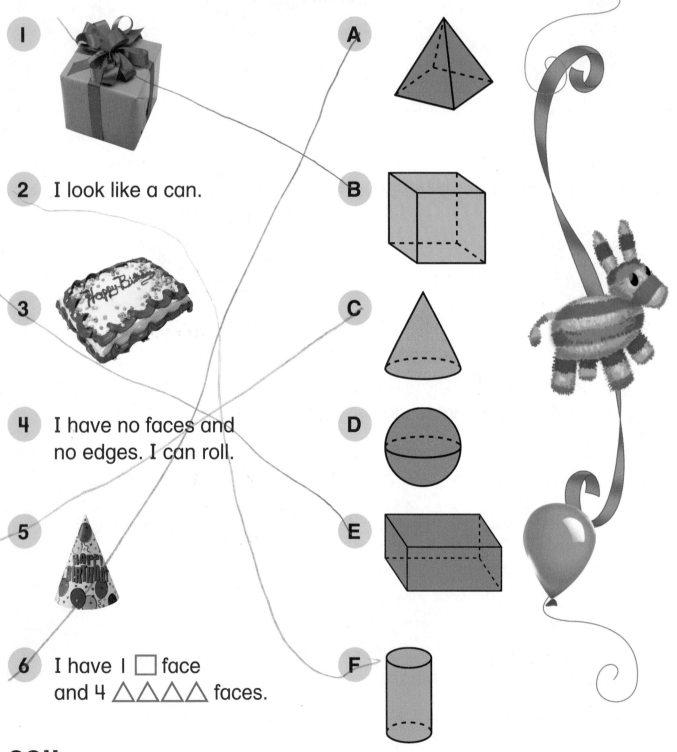

1

2 I look like a can.

3

4 I have no faces and no edges. I can roll.

5

6 I have 1 ☐ face and 4 △△△△ faces.

A

B

C

D

E

F

Name _____

Two-Dimensional Figures

Key Concept

Two-dimensional figures have length and width.

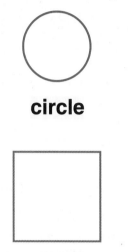

circle   **rectangle**

square   **triangle**

Vocabulary

circle   a closed, round, plane figure

rectangle   a plane figure with 4 sides and 4 corners

square   a rectangle that has 4 sides with the same length

triangle   a plane figure with three sides and three angles

The name of this musical instrument also describes its shape. It is a triangle.

Common objects can have the shape of a two-dimensional figure.

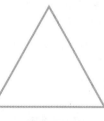

Example

Circle the similar figure.

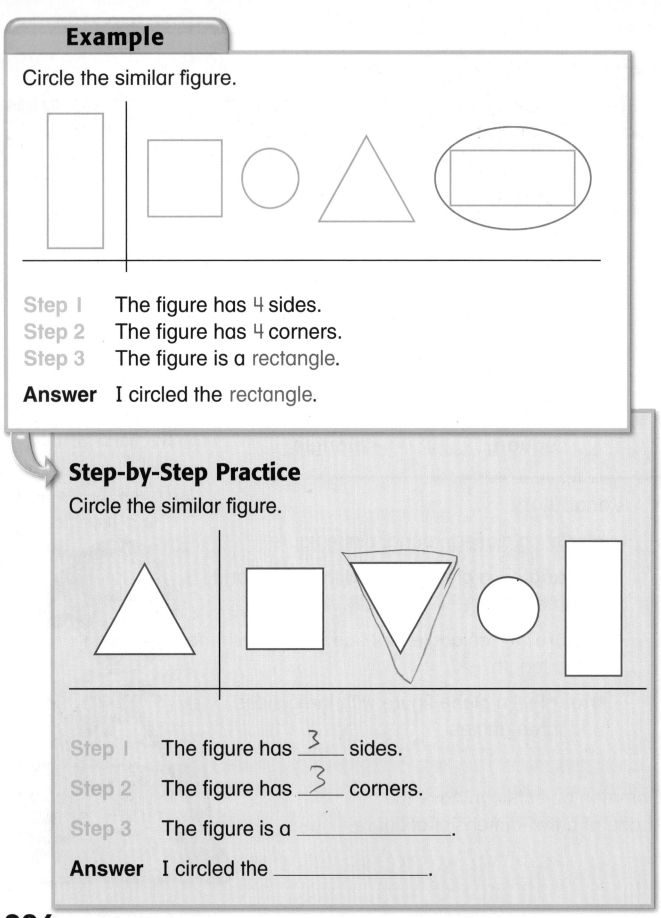

Step 1   The figure has 4 sides.
Step 2   The figure has 4 corners.
Step 3   The figure is a rectangle.

Answer   I circled the rectangle.

Step-by-Step Practice

Circle the similar figure.

Step 1   The figure has __3__ sides.

Step 2   The figure has __3__ corners.

Step 3   The figure is a _____.

Answer   I circled the _____.

Name _____

▶ Guided Practice

1 Circle the similar figure.

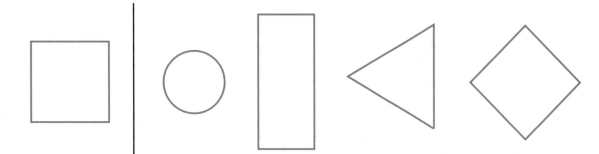

Problem-Solving Practice

2 Carmen drew two squares side by side so that they were touching. What new figure did Carmen draw?

Understand   Underline key words.

Plan   Draw a picture.

Solve   Draw two squares side by side.

The new figure is a _____.

Check   Outline the new figure you drew.

GO on

 **Practice on Your Own**

Circle the similar figure.

3
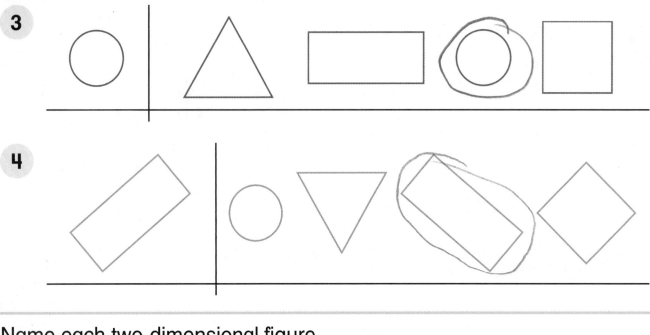

4

Name each two-dimensional figure.

5

rectangular

6

circle

7 WRITING IN ►MATH Andrew drew this picture of a house. What two-dimensional figures did he use to draw the house? Explain.

_____

_____

_____

Vocabulary Check   Complete.

8 A rectangle that has 4 sides with the same length

is a _____.

238 two hundred thirty-eight

Name _____

Sides and Vertices

Key Concept

This figure has 4 **sides**.

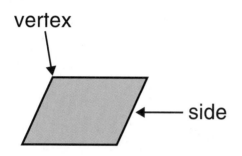

vertex

side

There are 4 points on this figure where the sides meet.
They are called **vertices**.

Vocabulary

side   one of the line segments that make up
a two-dimensional figure

vertex   a point where two or more sides meet;
more than 1 vertex are called vertices.

two-dimensional figure   a figure such as a
triangle or square that is flat

The number of sides is equal to the
number of vertices.

There are 8 vertices on a stop sign.
So, there are 8 sides on a stop sign.

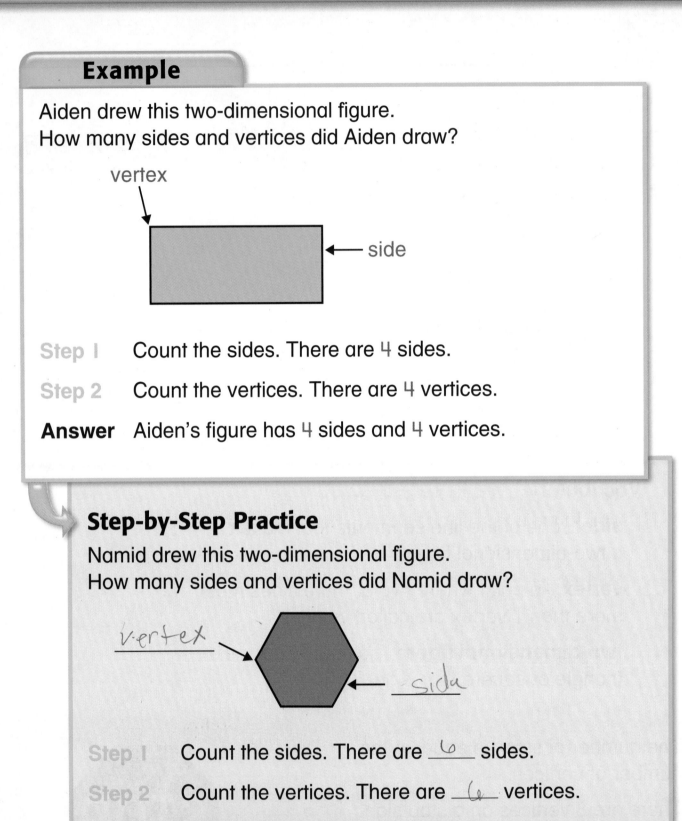

Example

Aiden drew this two-dimensional figure.
How many sides and vertices did Aiden draw?

vertex

side

Step 1   Count the sides. There are 4 sides.

Step 2   Count the vertices. There are 4 vertices.

Answer   Aiden's figure has 4 sides and 4 vertices.

Step-by-Step Practice

Namid drew this two-dimensional figure.
How many sides and vertices did Namid draw?

vertex

side

Step 1   Count the sides. There are __6__ sides.

Step 2   Count the vertices. There are __6__ vertices.

Answer   Namid's figure has __6__ sides

and __6__ vertices.

Name _____

▶ Guided Practice

Count the sides and vertices of each figure.

1

_3_ sides

_3_ vertices

2

_4_ sides

_4_ vertices

Problem-Solving Practice

3 Jamal found these two puzzle pieces. The missing piece to his rocket puzzle has 10 sides. Which piece is missing from Jamal's rocket puzzle?

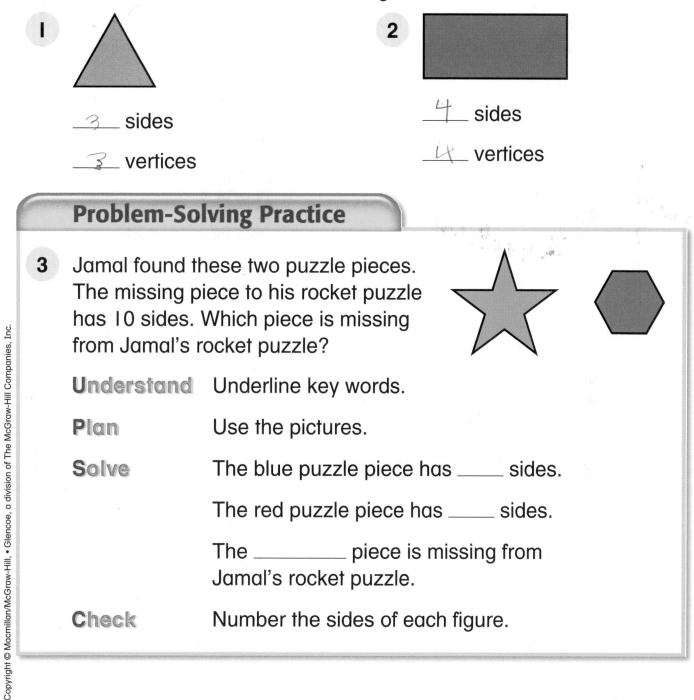

Understand   Underline key words.

Plan   Use the pictures.

Solve   The blue puzzle piece has _____ sides.

The red puzzle piece has _____ sides.

The _____ piece is missing from Jamal's rocket puzzle.

Check   Number the sides of each figure.

GO on

▶ Practice on Your Own

Count the sides and vertices of each figure.

4

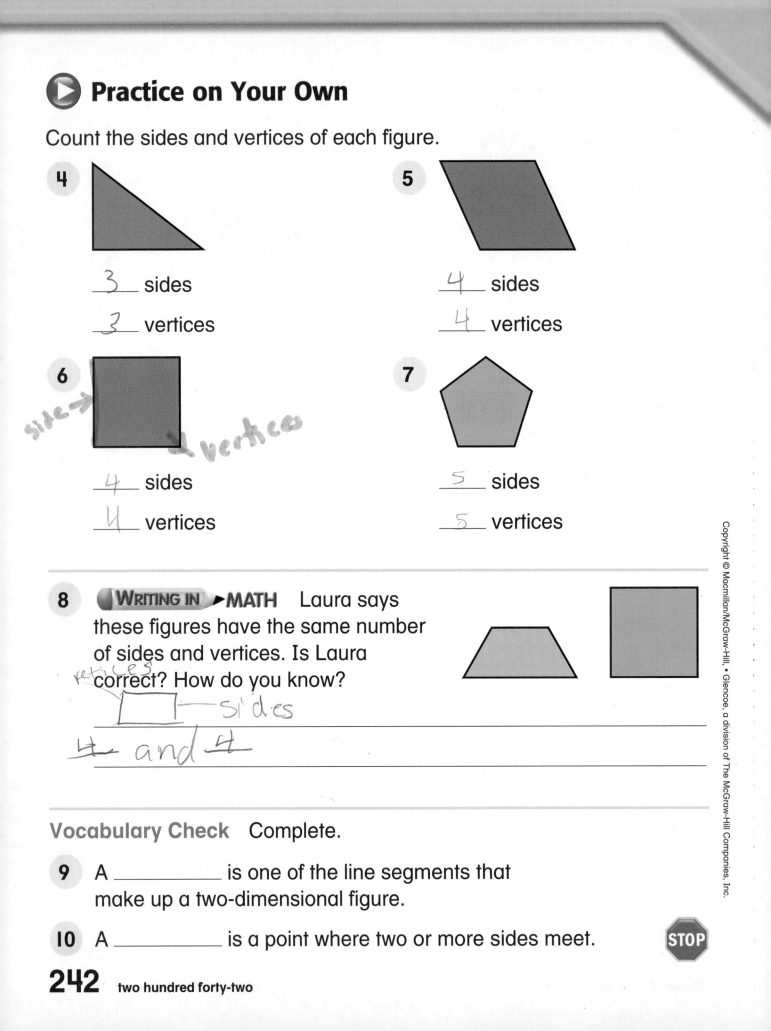

_3_ sides

_3_ vertices

5

_4_ sides

_4_ vertices

6

side→

2 vertices

_4_ sides

_4_ vertices

7

_5_ sides

_5_ vertices

8 **WRITING IN ▶MATH** Laura says these figures have the same number of sides and vertices. Is Laura correct? How do you know?

vertices

☐ —sides

4 and 4

Vocabulary Check   Complete.

9 A _____ is one of the line segments that make up a two-dimensional figure.

10 A _____ is a point where two or more sides meet.

STOP

242   two hundred forty-two

Name _____

Relate Two- and Three-Dimensional Figures

Key Concept

Figure	Faces	Vertices
cube	6 ▢	8
pyramid	1 ▢ and 4 △	5
rectangular prism	6 ▭	8
cone	1 ○	1
sphere	no faces	0

The faces of a rectangular prism are rectangles.

A square is a two-dimensional figure. It has length and width.

A cube is a three-dimensional figure. It has length, width, and height.

Some three-dimensional figures have faces that are two-dimensional figures.

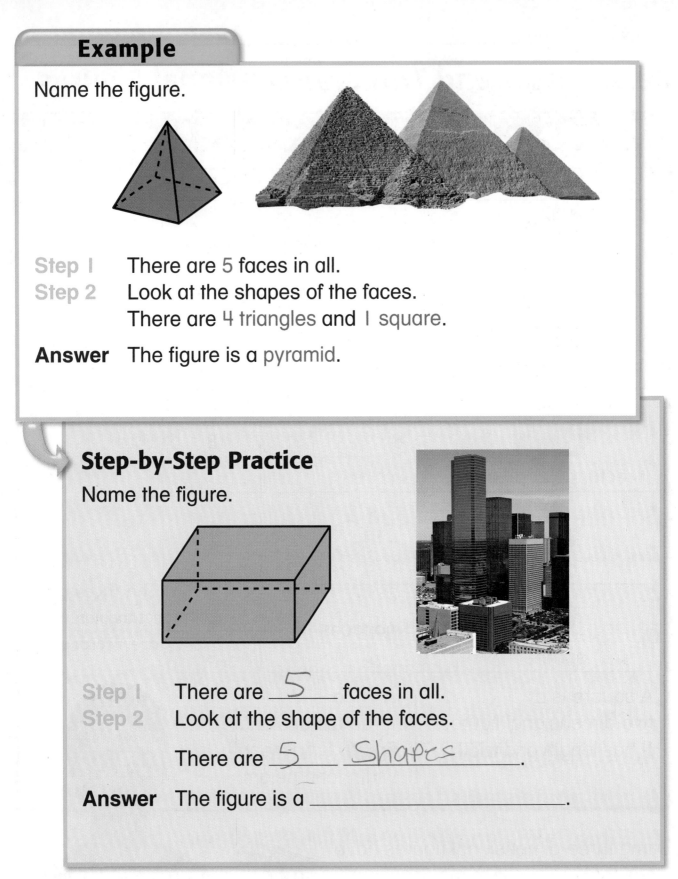

Example

Name the figure.

Step 1 There are 5 faces in all.
Step 2 Look at the shapes of the faces.
There are 4 triangles and 1 square.

Answer The figure is a pyramid.

Step-by-Step Practice

Name the figure.

Step 1 There are ___5___ faces in all.
Step 2 Look at the shape of the faces.

There are __5__ _Shapes_ .

Answer The figure is a _____ .

Name _____

▶ Guided Practice

Write the number of faces for each figure.

1 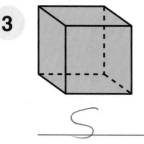 ___2___ faces

2  ___5___ faces

Name the shape of the faces for each figure.

3 ___5___

4 ___5___

Problem-Solving Practice

5 Olivia wants to make a pyramid. What two-dimensional figures will she need?

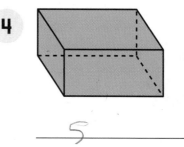

Understand   Underline key words.

Plan   Use a model.

Solve   Look at the faces of a pyramid.

Olivia will need _____ _____

and _____ _____.

Check   Did you count all the faces on the pyramid?

GO on

▶ Practice on Your Own

Write the number of faces for each figure.

6
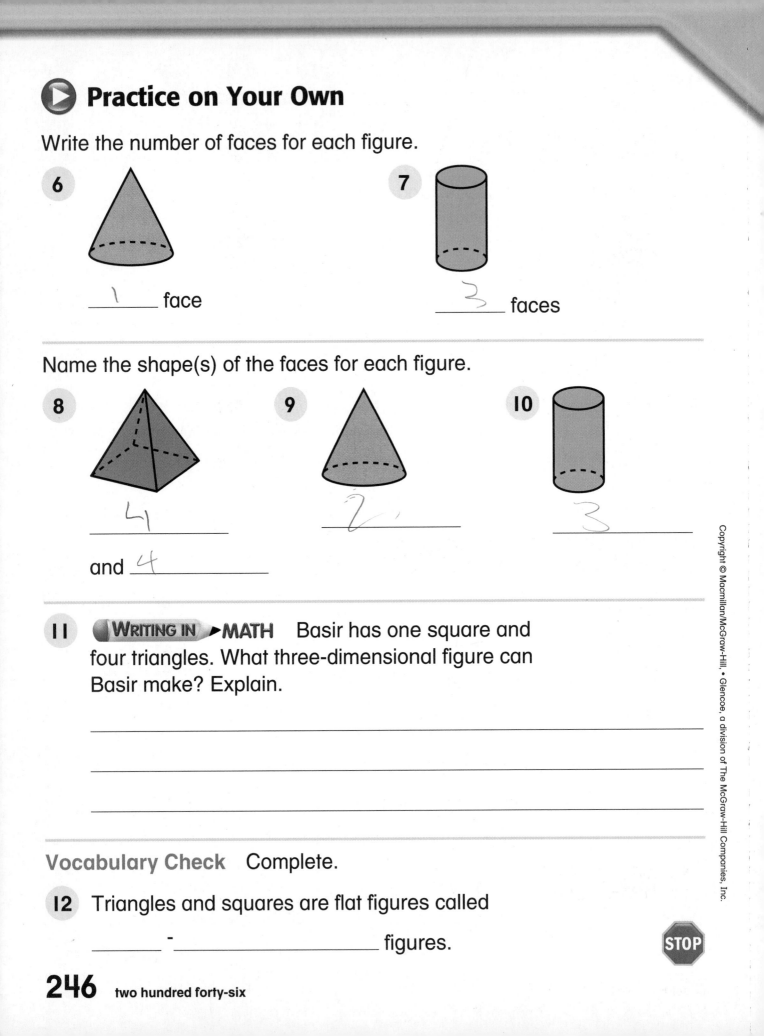
_____1_____ face

7
_____3_____ faces

Name the shape(s) of the faces for each figure.

8
_____4_____

and _____4_____

9
_____2_____

10
_____3_____

---

11 **WRITING IN ▶MATH**   Basir has one square and four triangles. What three-dimensional figure can Basir make? Explain.

_____

_____

_____

Vocabulary Check   Complete.

12 Triangles and squares are flat figures called

_____-_____ figures.

STOP

Name _____

Progress Check 2 (Lessons 7-3, 7-4, and 7-5)

Count the sides and vertices of each figure.

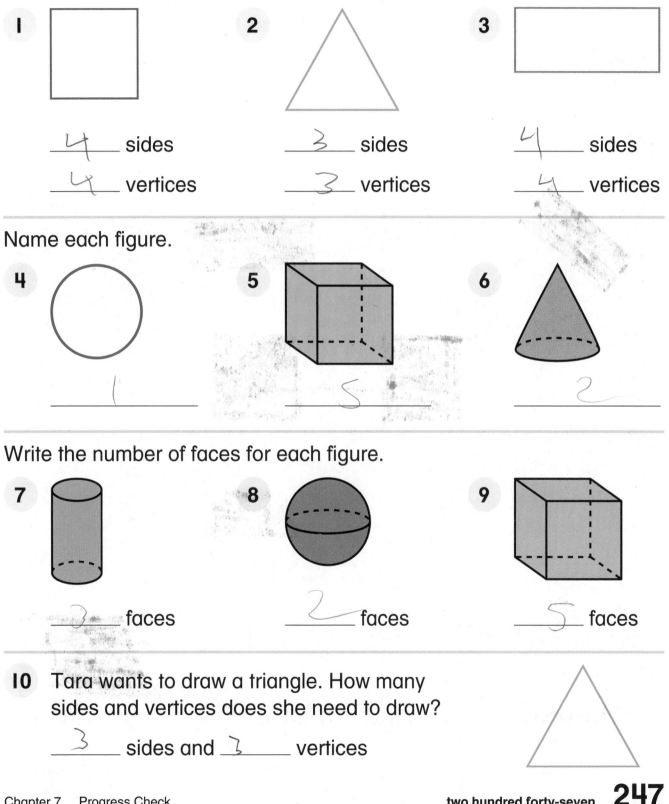

1

__4__ sides

__4__ vertices

2

__3__ sides

__3__ vertices

3

__4__ sides

__4__ vertices

Name each figure.

4

_____1_____

5

_____5_____

6

_____2_____

Write the number of faces for each figure.

7

__3__ faces

8

__2__ faces

9

__5__ faces

10 Tara wants to draw a triangle. How many
sides and vertices does she need to draw?

__3__ sides and __3__ vertices

Name _____

Help Anna the Ant find her way to the ant hill!

Shade the boxes that show a two-dimensional figure with 4 sides and 4 vertices.

Connect four in a row to show Anna the Ant which way to go.

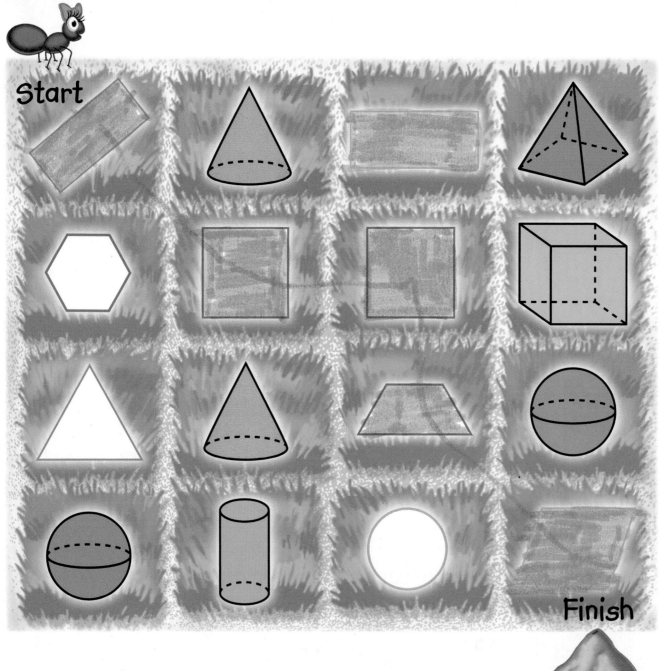

Name _____

Review

Vocabulary

Word Bank

edge ✓

face ✓

side ✓

vertex ✓

Use the Word Bank to complete.

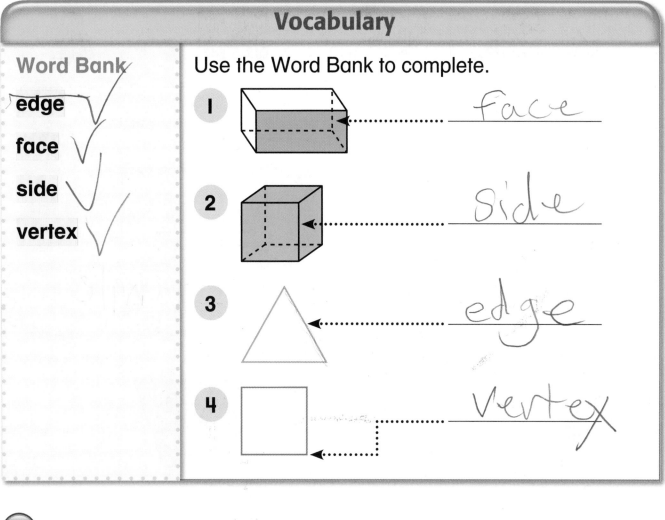

1   face

2   side

3   edge

4   vertex

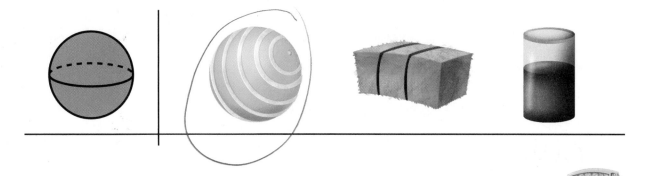

▶ Concepts

5 Circle the object that is similar to the figure.

6 Circle the figure that is similar to the blue face.

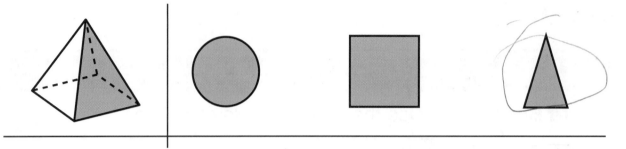

Name each two-dimensional figure.

7

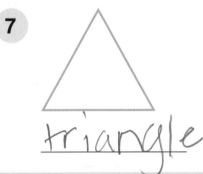

triangle

8

Rectangle

Count the sides and vertices of each figure.

9

_3_ sides

_3_ vertices

10

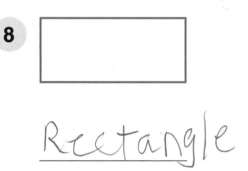

_4_ sides

_4_ vertices

Name the shape(s) of the faces for each figure.

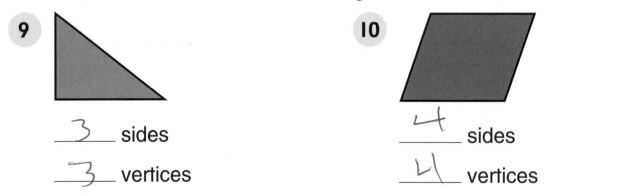

11

cube

12

Pyramid

and _____

13

Rectangle

Prism

Review

STOP

Name _____

Chapter Test

Name each figure.

1 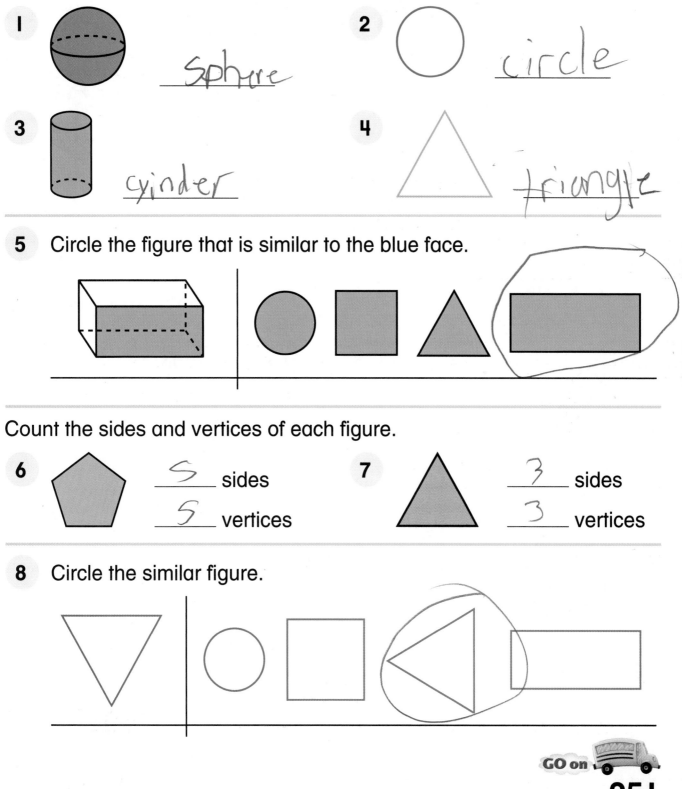 Sphere

2 circle

3 cyinder

4 triangle

5 Circle the figure that is similar to the blue face.

Count the sides and vertices of each figure.

6 _5_ sides
   _5_ vertices

7 _3_ sides
   _3_ vertices

8 Circle the similar figure.

GO on

9 Who is Correct?

Mike and Benito describe a cube.

A cube has 6 faces.

Mike

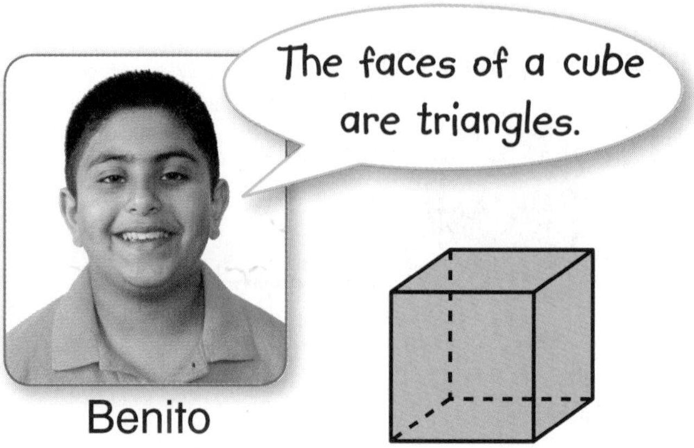

The faces of a cube are triangles.

Benito

Circle the correct answer. Explain.

Mike is right.

Write the number and shape(s) of the faces for each figure.

10

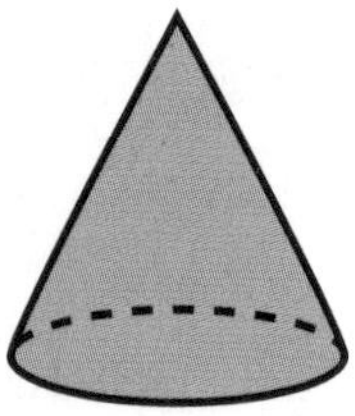

___2___ _____ faces

11

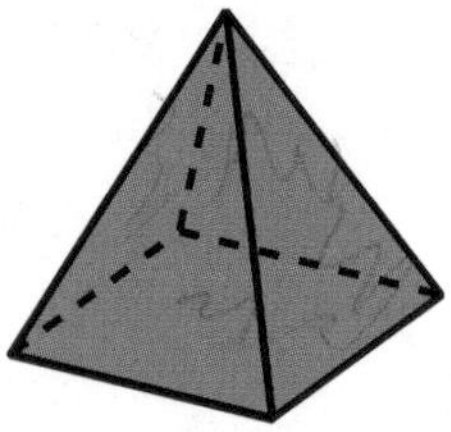

___4___ _____ face and

_____ _____ faces

12 Charlotte's mom packs a juice box with her lunch. What three-dimensional figure is similar to the juice box?

_____

Test Practice

Choose the correct answer.

1 Kylie describes a three-dimensional figure. She says the figure has 1 square face and 4 triangle faces. What figure does Kylie describe?

- ○ pyramid
- ○ rectangular prism
- ○ triangle
- ○ cylinder

2 Name the two-dimensional figure.

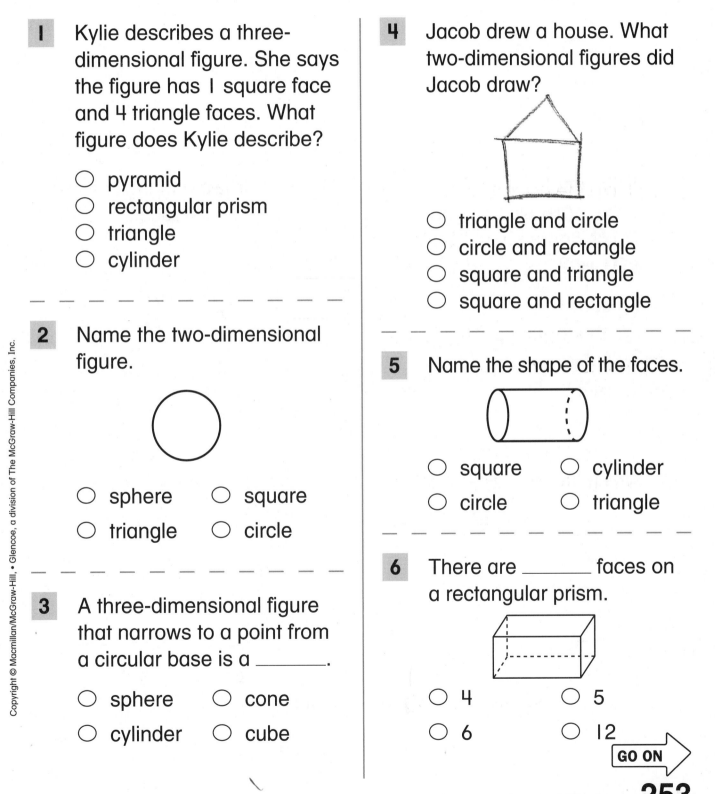

- ○ sphere   ○ square
- ○ triangle   ○ circle

3 A three-dimensional figure that narrows to a point from a circular base is a _____.

- ○ sphere   ○ cone
- ○ cylinder   ○ cube

4 Jacob drew a house. What two-dimensional figures did Jacob draw?

- ○ triangle and circle
- ○ circle and rectangle
- ○ square and triangle
- ○ square and rectangle

5 Name the shape of the faces.

- ○ square   ○ cylinder
- ○ circle   ○ triangle

6 There are _____ faces on a rectangular prism.

- ○ 4   ○ 5
- ○ 6   ○ 12

GO ON

7 Lucy draws a pyramid. How many edges does Lucy draw?

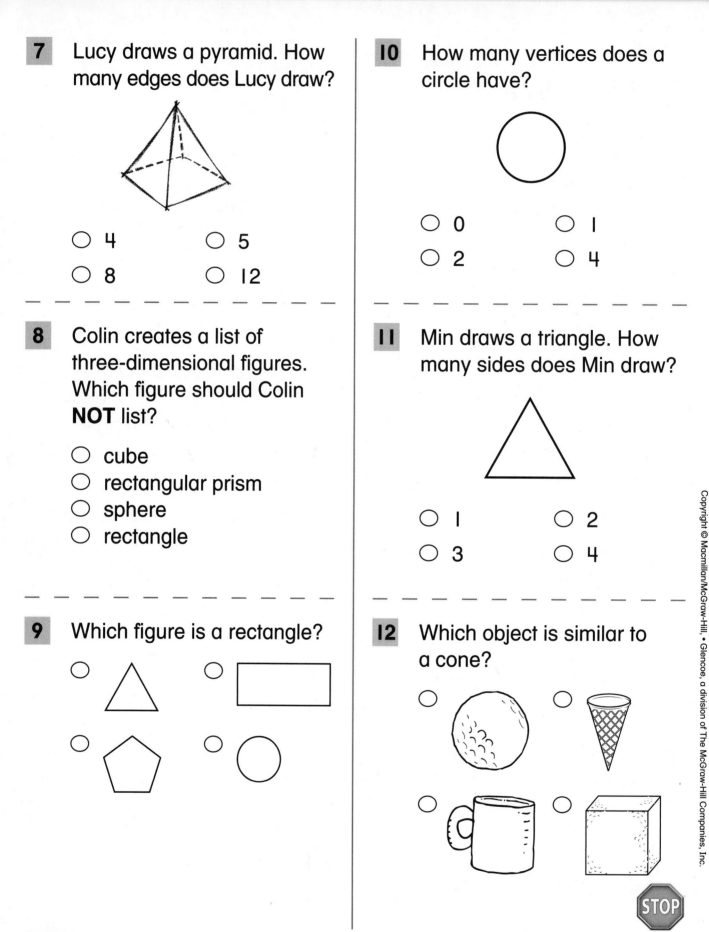

- ○ 4
- ○ 5
- ○ 8
- ○ 12

8 Colin creates a list of three-dimensional figures. Which figure should Colin **NOT** list?

- ○ cube
- ○ rectangular prism
- ○ sphere
- ○ rectangle

9 Which figure is a rectangle?

- ○
- ○
- ○
- ○

10 How many vertices does a circle have?

- ○ 0
- ○ 1
- ○ 2
- ○ 4

11 Min draws a triangle. How many sides does Min draw?

- ○ 1
- ○ 2
- ○ 3
- ○ 4

12 Which object is similar to a cone?

- ○
- ○
- ○
- ○

STOP

Test Practice

Home Connection

English

Spanish

Dear Family,

Today our class started **Chapter 8, Measurement and Geometry.** In this chapter, I will compare the sizes and shapes of figures. I will also compose and decompose figures and determine symmetry and congruence.

Love, _____

Estimada familia:

Hoy en clase comenzamos el **Capítulo 8, Medición y Geometría.** En este capítulo compararé los tamaños y formas de las figuras. También formaré y separaré las partes de figuras, y determinaré su simetría y su congruencia.

Cariños, _____

Help at Home

You can help your child by asking him or her to compare the sizes and shapes of common household objects. You can also do paper folding activities to show symmetry.

Ayude en casa

Usted puede ayudar a su hijo(a) pidiéndole que compare las formas y el tamaño de objetos comunes del hogar. También puede hacer actividades en las que doble papel para demostrar la simetría de las figuras.

Math Online ▸ Take the chapter Get Ready quiz at macmillanmh.com.

Name _____

Get Ready

Circle the greater number.

1 5 **9**

2 **4** 2

Count each number of sides.

3

4 sides

4

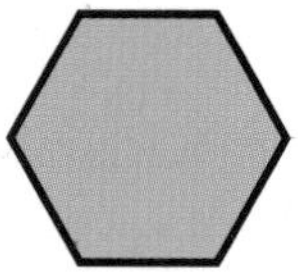

6 sides

5

3 sides

Name each shape.

6

_____

7

cube

8

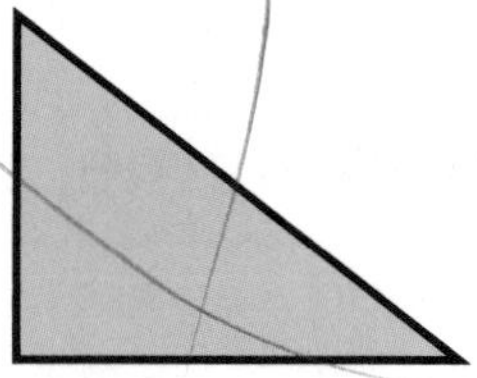

_____

9

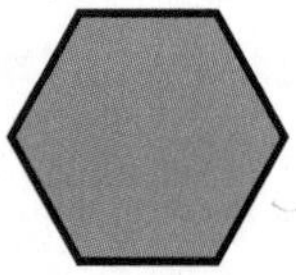

_____

10

rectangle

11

circle

STOP

Name _____

Compare Size

Key Concept

You can **compare** the size of two figures.

These figures are the same size.   These figures are different sizes.

Copyright © Macmillan/McGraw-Hill, • Glencoe, a division of The McGraw-Hill Companies, Inc.

Vocabulary

compare   look at objects, shapes, or numbers and see how they are the same and different

Count how many units tall and wide figures are to compare size.

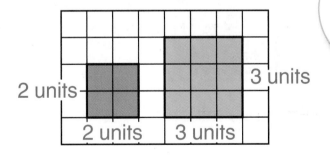

2 units

2 units   3 units

3 units

The red square is 2 units tall and 2 units wide. The green square is 3 units tall and 3 units wide. The squares are different sizes.

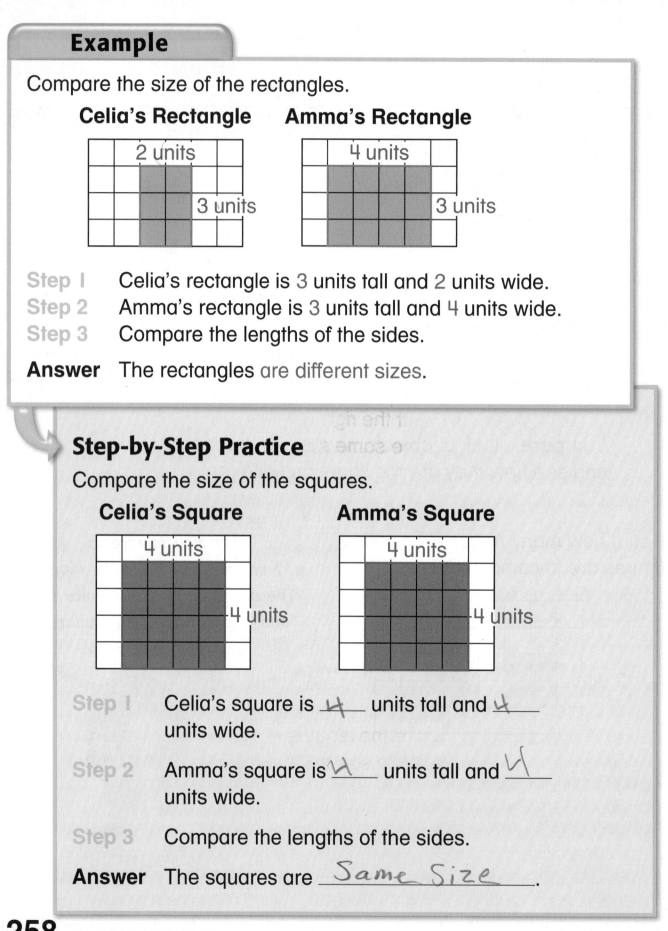

Example

Compare the size of the rectangles.

Celia's Rectangle   **Amma's Rectangle**

2 units
3 units

4 units
3 units

Step 1   Celia's rectangle is 3 units tall and 2 units wide.
Step 2   Amma's rectangle is 3 units tall and 4 units wide.
Step 3   Compare the lengths of the sides.

Answer   The rectangles are different sizes.

Step-by-Step Practice

Compare the size of the squares.

Celia's Square   **Amma's Square**

4 units
4 units

4 units
4 units

Step 1   Celia's square is __4__ units tall and __4__ units wide.

Step 2   Amma's square is __4__ units tall and __4__ units wide.

Step 3   Compare the lengths of the sides.

Answer   The squares are ___Same Size___.

Name _____

▶ Guided Practice

Compare the size. Write "same" or "different."

1

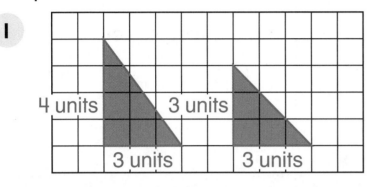

The triangles are

Same sizes.

Problem-Solving Practice

2 Keith drew the square at the right.
Draw a square that is the same size
as Keith's square.

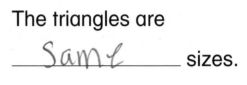

2 units

2 units

Understand   Underline key words.

Plan   Draw a picture on a grid.

Solve

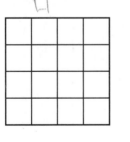

Keith's rectangle is

_4_ units tall and

_4_ units wide.

Check   Compare the lengths of the sides of your square
and Keith's square. Are they the same?

GO on 🚌

▶ Practice on Your Own

3 Compare the size. Write "same" or "different."

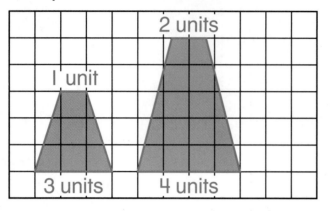

The trapezoids are

different sizes.

---

4 Draw a square that is the same size as the one shown.

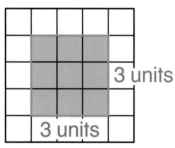 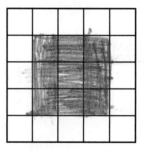

---

5 WRITING IN ▶MATH Are the figures the same size or different sizes? Explain.

_____

_____

_____

_____

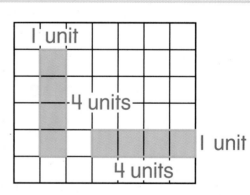

Vocabulary Check Complete.

6 When you _____ figures, you look to see how they are the same and different.

STOP

260 two hundred sixty

Name _____

Compare Shape

Key Concept

You can **compare** the shape of two figures.

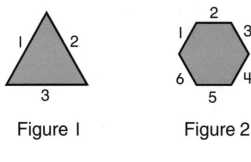

Figure 1   Figure 2

3 sides   6 sides

The figures do not look the same.
Figure 1 and Figure 2 are different shapes.

Vocabulary

compare   look at objects, shapes, or numbers and see how they are alike and different

side   one of the line segments that make up a shape

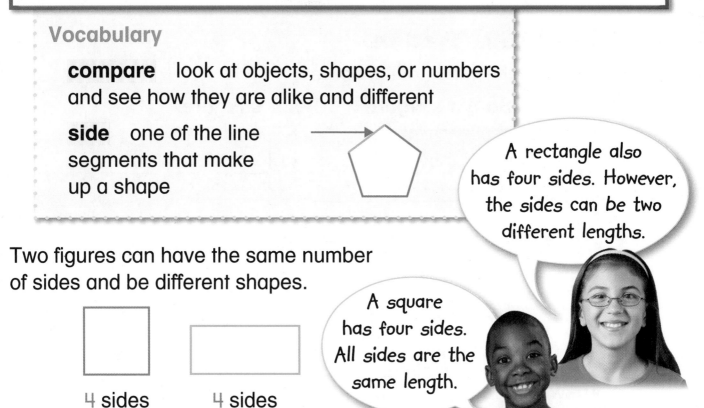

Two figures can have the same number of sides and be different shapes.

4 sides   4 sides

A rectangle also has four sides. However, the sides can be two different lengths.

A square has four sides. All sides are the same length.

Example

Compare the shape of the figures.

Kendra's Figure   **Simone's Figure**

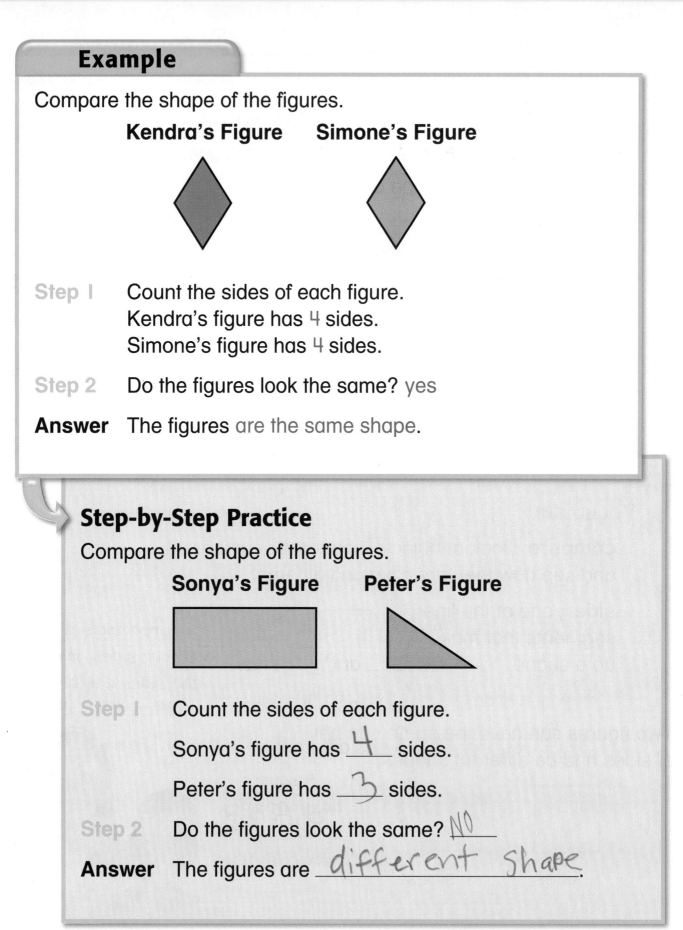

Step 1   Count the sides of each figure.
Kendra's figure has 4 sides.
Simone's figure has 4 sides.

Step 2   Do the figures look the same? yes

Answer   The figures are the same shape.

Step-by-Step Practice

Compare the shape of the figures.

Sonya's Figure   **Peter's Figure**

Step 1   Count the sides of each figure.

Sonya's figure has __4__ sides.

Peter's figure has __3__ sides.

Step 2   Do the figures look the same? __NO__

Answer   The figures are __different shape__.

Name _____

▶ Guided Practice

1 Compare the shape. Write "same" or "different."

Same

Problem-Solving Practice

2 Mrs. Perry drew three figures on the board.

Figure 1 **Figure 2** **Figure 3**

Which figure is a different shape?

Understand   Underline key words.

Plan   Compare the figures.

Solve   Count the sides of each figure.

Figures __1__ and __3__ have __3__ sides.

Figure __2__ has __4__ sides.

Figure __2__ is different.

Check   Does the figure you chose look different than the others? _different side_

GO on

▶ Practice on Your Own

Compare the shape. Write "same" or "different."

3 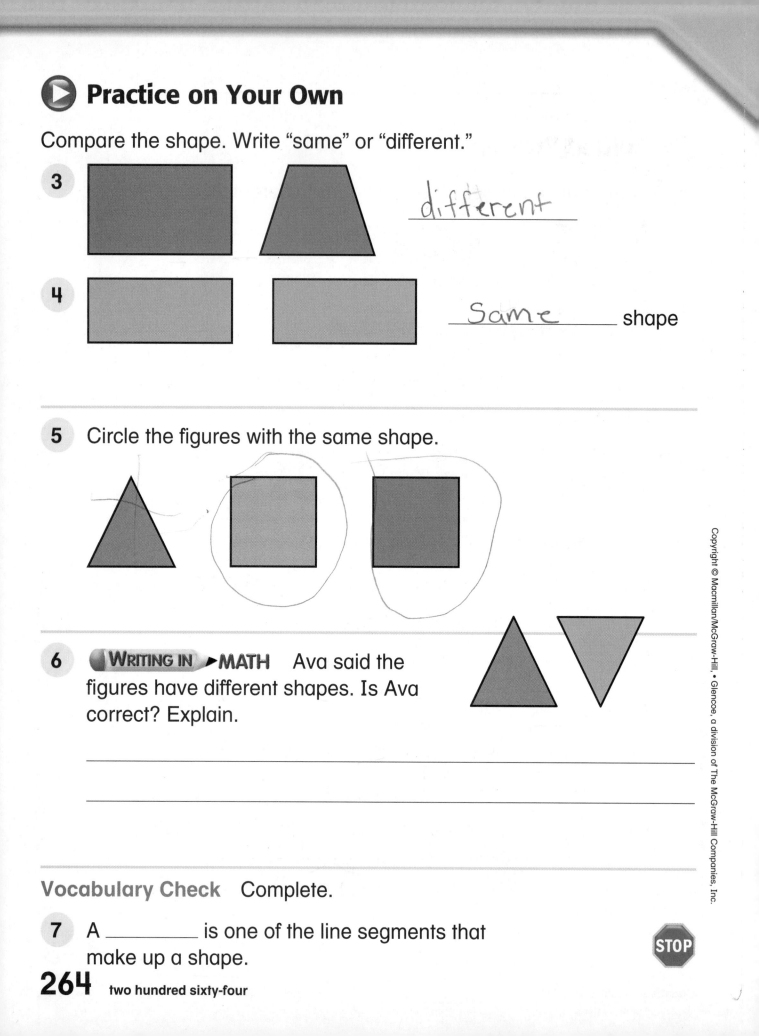 _different_

4 _Same_ shape

---

5 Circle the figures with the same shape.

6 **WRITING IN ►MATH** Ava said the figures have different shapes. Is Ava correct? Explain.

_____

_____

Vocabulary Check Complete.

7 A _____ is one of the line segments that make up a shape.

264 two hundred sixty-four

Name _____

Progress Check 1 (Lessons 8-1 and 8-2)

Compare the size. Write "same" or "different."

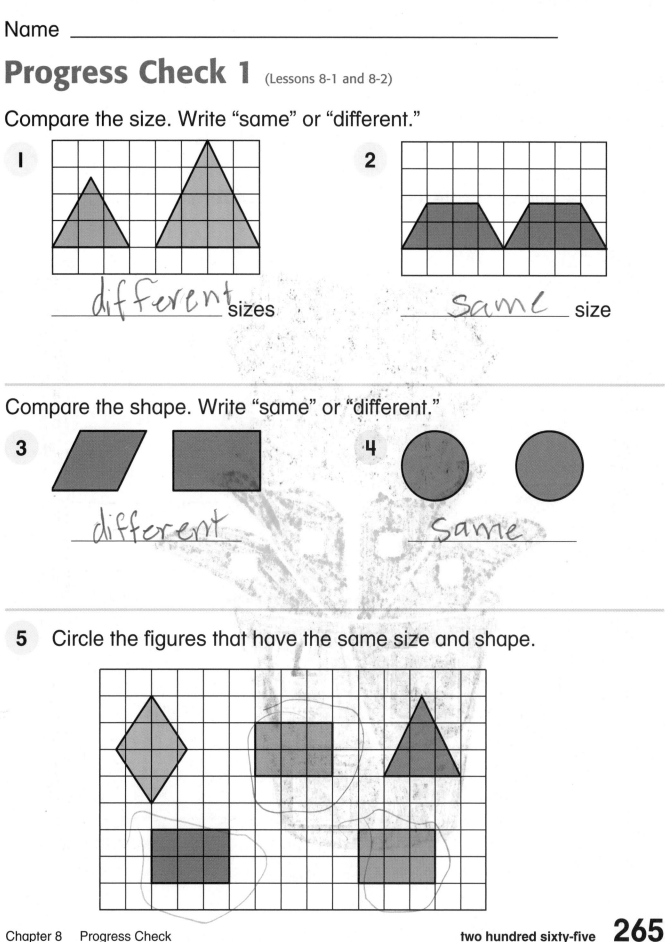

1 ___different___ sizes

2 ___same___ size

Compare the shape. Write "same" or "different."

3 ___different___

4 ___same___

5 Circle the figures that have the same size and shape.

Name _____

Color by Figure

Compare the figures shown in the picture size and shape.
Color the sections **green** for figures that are the same.
Color the sections **orange** for figures that are **different sizes**.
Color the sections **purple** for figures that are **different shapes**.

Name _____

Create Figures

Key Concept

You can put figures together to **create** a new figure.

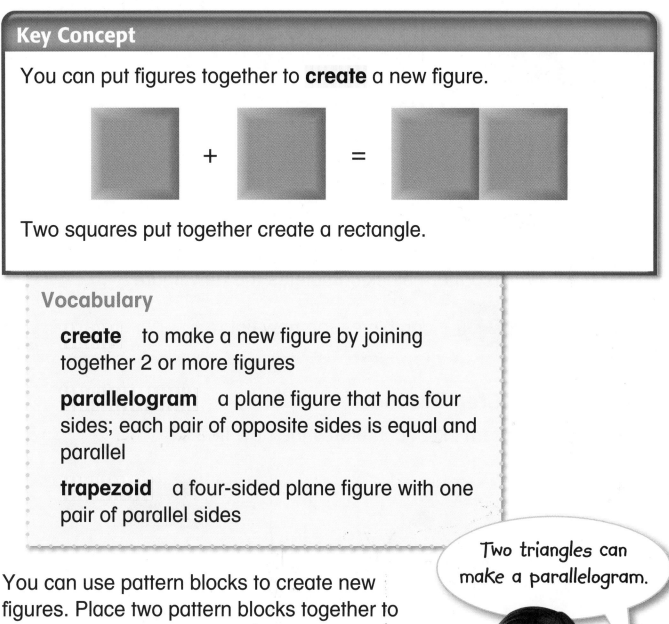

Two squares put together create a rectangle.

Vocabulary

create   to make a new figure by joining together 2 or more figures

parallelogram   a plane figure that has four sides; each pair of opposite sides is equal and parallel

trapezoid   a four-sided plane figure with one pair of parallel sides

You can use pattern blocks to create new figures. Place two pattern blocks together to create a new figure.

Two triangles can make a parallelogram.

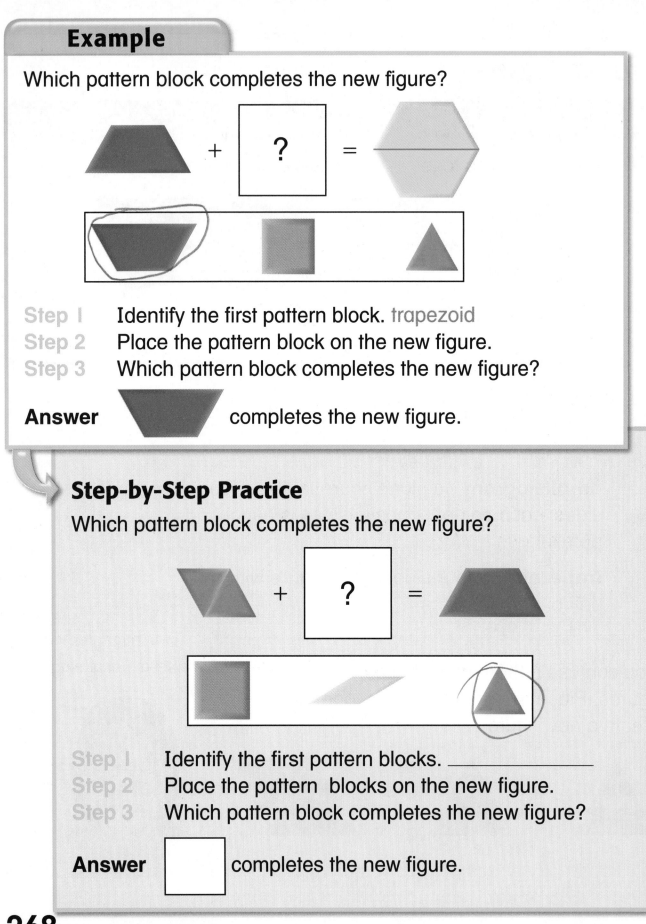

Example

Which pattern block completes the new figure?

Step 1 Identify the first pattern block. trapezoid
Step 2 Place the pattern block on the new figure.
Step 3 Which pattern block completes the new figure?

Answer _____ completes the new figure.

Step-by-Step Practice

Which pattern block completes the new figure?

Step 1 Identify the first pattern blocks. _____
Step 2 Place the pattern blocks on the new figure.
Step 3 Which pattern block completes the new figure?

Answer [ ] completes the new figure.

Name _____

▶ Guided Practice

1   Circle the pattern block that completes the new figure.

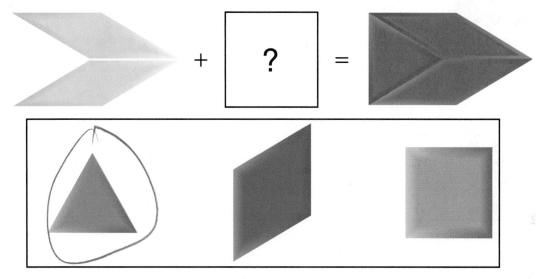

Problem-Solving Practice

2   Matias wants to use two pattern blocks to
    make a house. What blocks can he use?

Understand	Underline key words.
Plan	Act it out.
Solve	Use pattern blocks to make a house.

Matias can use a _____ and

a _____ to make a house.

Check	Did you use two pattern blocks to make a house?

GO on

Practice on Your Own

3 Circle the pattern block that completes the new figure.

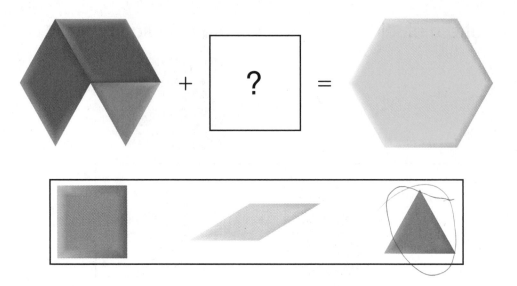

---

4 WRITING IN ▸MATH   Lilly wants to make a parallelogram using two trapezoids. Can she do it? Explain.

_____

_____

_____

Vocabulary Check   Complete.

5 I can _____ a new figure by joining together 2 or more figures.

270   two hundred seventy

Name _____

Take Apart Figures

Key Concept

You can draw a line through a figure to make two new figures.

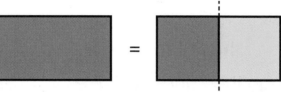

This **rectangle** can be taken apart, or split, into two **squares**.

Vocabulary

rectangle   a plane figure with four sides and four corners

square   a rectangle that has four equal sides

parallelogram   a plane figure that has four sides; each pair of opposite sides is equal and parallel

trapezoid   a four-sided plane figure with only two opposite sides that are the same length

I can take apart a hexagon into two trapezoids.

A rectangle can be taken apart into more than two figures.

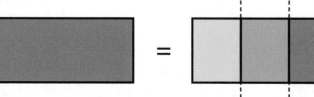

You can draw two lines to make three new rectangles.

Draw one line to take apart the diamond into two triangles.

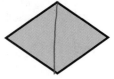

Step 1   Look at the figure.

Step 2   Identify a triangle in the figure.

Step 3   Draw a line through the figure to make two triangles.

Answer

Step-by-Step Practice

Draw one line to take apart the parallelogram into two trapezoids.

Step 1   Look at the figure.

Step 2   Identify a trapezoid in the figure.

Step 3   Draw a line through the figure to make two trapezoids.

Answer

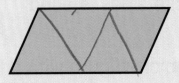

Name _____

▶ Guided Practice

Draw one line to make the given figures.

1 2 triangles

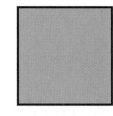

2 2 trapezoids

Problem-Solving Practice

3 Alex drew two lines through a trapezoid to make three new figures. One of the figures is a rectangle. What are the other two figures?

Understand   Underline key words.

Plan   Draw a picture.

Solve   Draw two lines to make one rectangle.

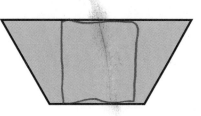

The other two figures are _____.

Check   Is one of the figures you created a rectangle?

GO on

▶ Practice on Your Own

Draw one line to make the given figures.

4 I parallelogram and I triangle

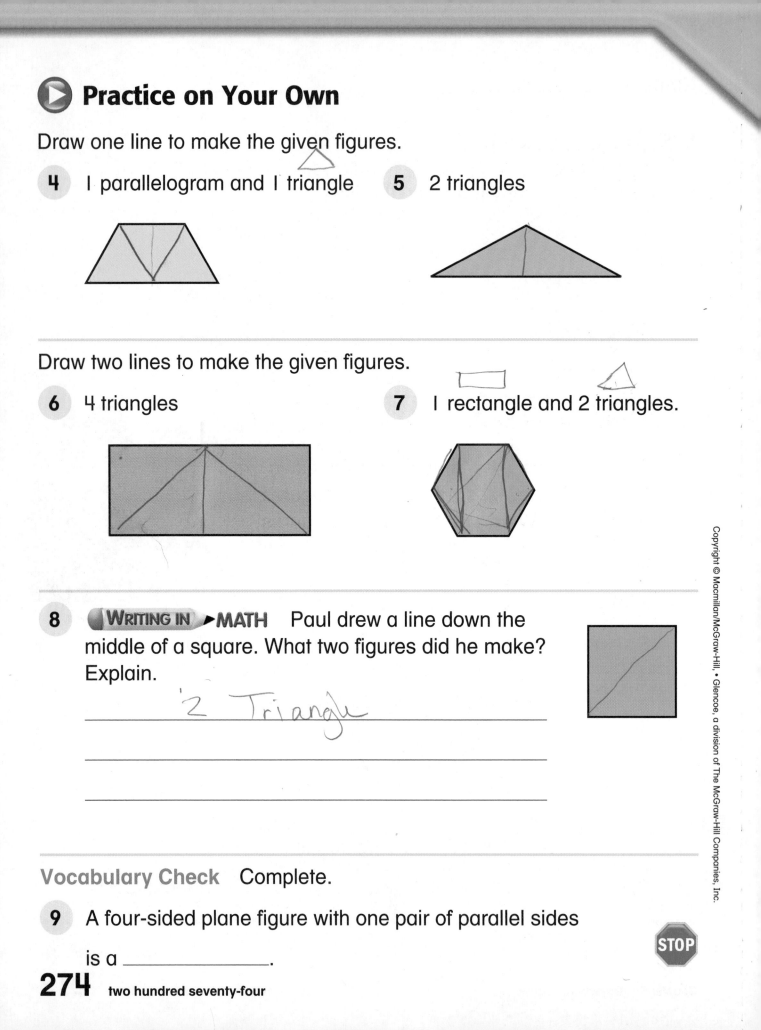

5 2 triangles

Draw two lines to make the given figures.

6 4 triangles

7 I rectangle and 2 triangles.

8 **WRITING IN ►MATH** Paul drew a line down the middle of a square. What two figures did he make? Explain.

2 Triangle

Vocabulary Check Complete.

9 A four-sided plane figure with one pair of parallel sides

is a _____.

Name _____

Progress Check 2 (Lessons 8-3 and 8-4)

1 Circle the pattern block that completes the new figure.

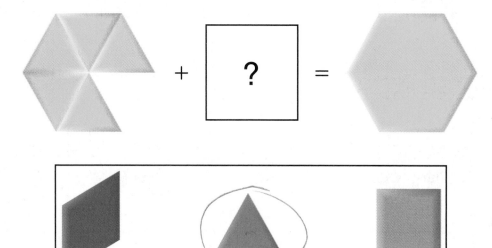

---

2 Draw the new figure.

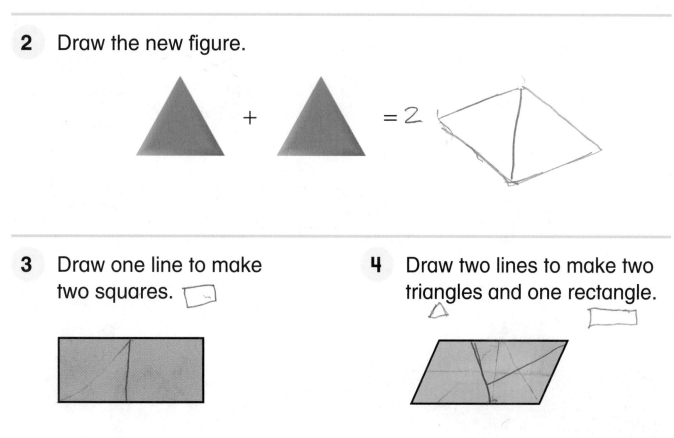

---

3 Draw one line to make two squares.

4 Draw two lines to make two triangles and one rectangle.

Name _____

Rabbit Maze

Help Rascal the Rabbit find his way through the maze.
Move from one square to the next by creating and taking apart figures.
You may not go through all of the squares.

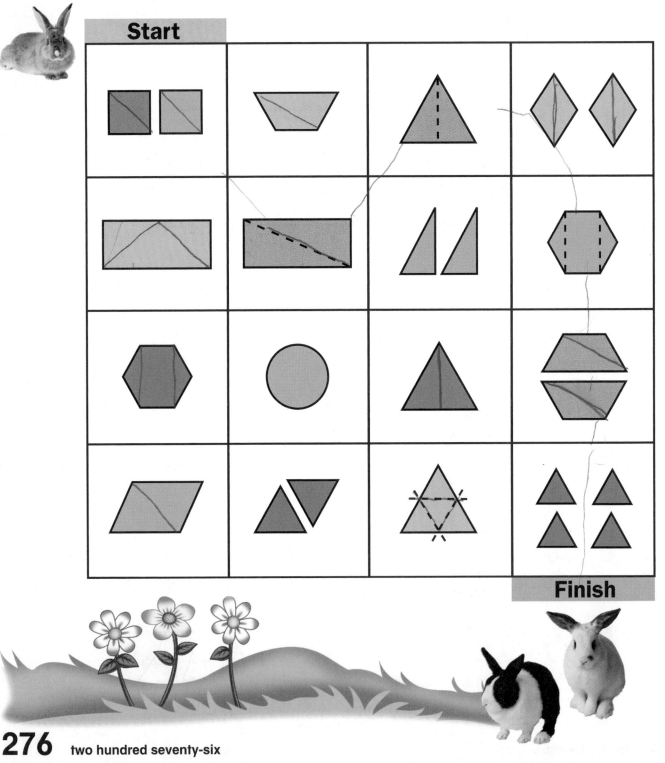

Name _____

Congruence

Key Concept

You can **compare** the figures below.

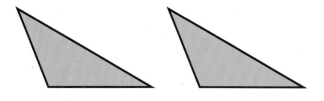

The figures are the same shape.
The figures are the same size.
The figures are **congruent**.

Vocabulary

compare   look at objects, shapes, or numbers and see how they are alike and different

congruent   figures that are the same shape and size

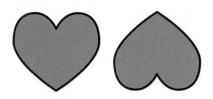

The hearts are congruent.

Figures in different positions are still congruent if they are the same shape and size.

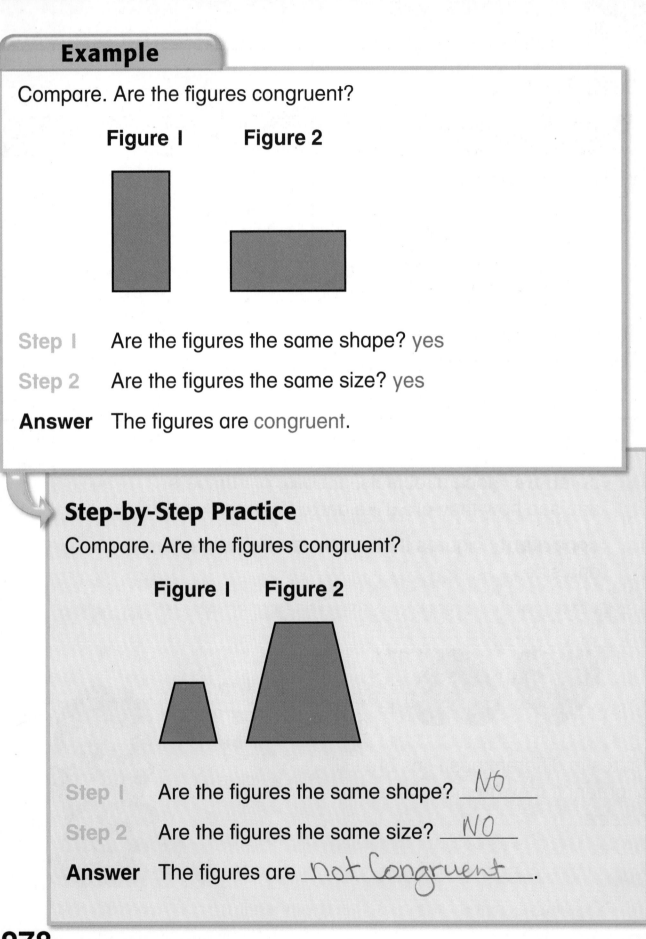

Example

Compare. Are the figures congruent?

Figure 1   **Figure 2**

Step 1   Are the figures the same shape? yes

Step 2   Are the figures the same size? yes

Answer   The figures are congruent.

Step-by-Step Practice

Compare. Are the figures congruent?

Figure 1   **Figure 2**

Step 1   Are the figures the same shape? _NO_

Step 2   Are the figures the same size? _NO_

Answer   The figures are _not Congruent_.

▶ Guided Practice

Compare. Are the figures congruent? Write "yes" or "no."

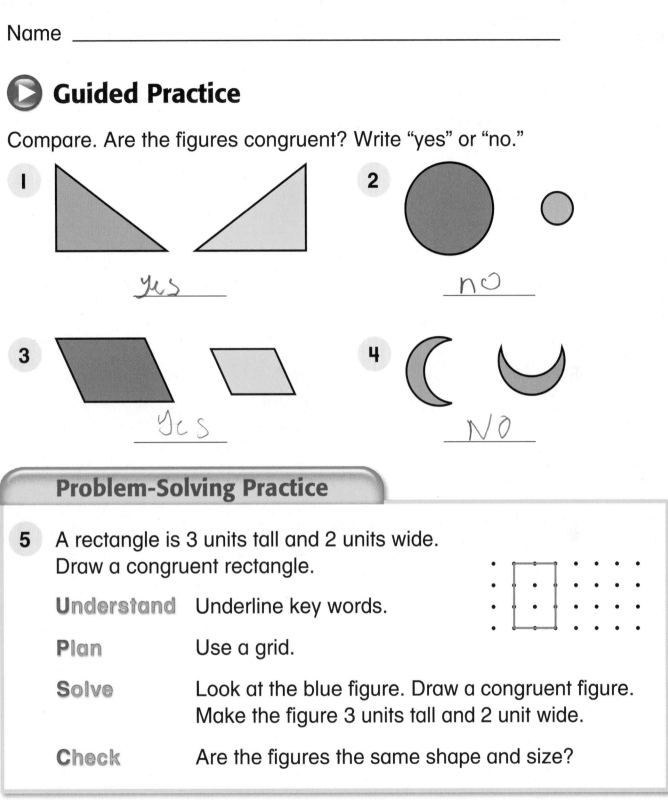

1. _____yes_____

2. _____no_____

3. _____yes_____

4. _____NO_____

Problem-Solving Practice

5   A rectangle is 3 units tall and 2 units wide.
    Draw a congruent rectangle.

    Understand   Underline key words.

    Plan   Use a grid.

    Solve   Look at the blue figure. Draw a congruent figure.
    Make the figure 3 units tall and 2 unit wide.

    Check   Are the figures the same shape and size?

GO on

▶ Practice on Your Own

Compare. Are the figures congruent? Write "yes" or "no."

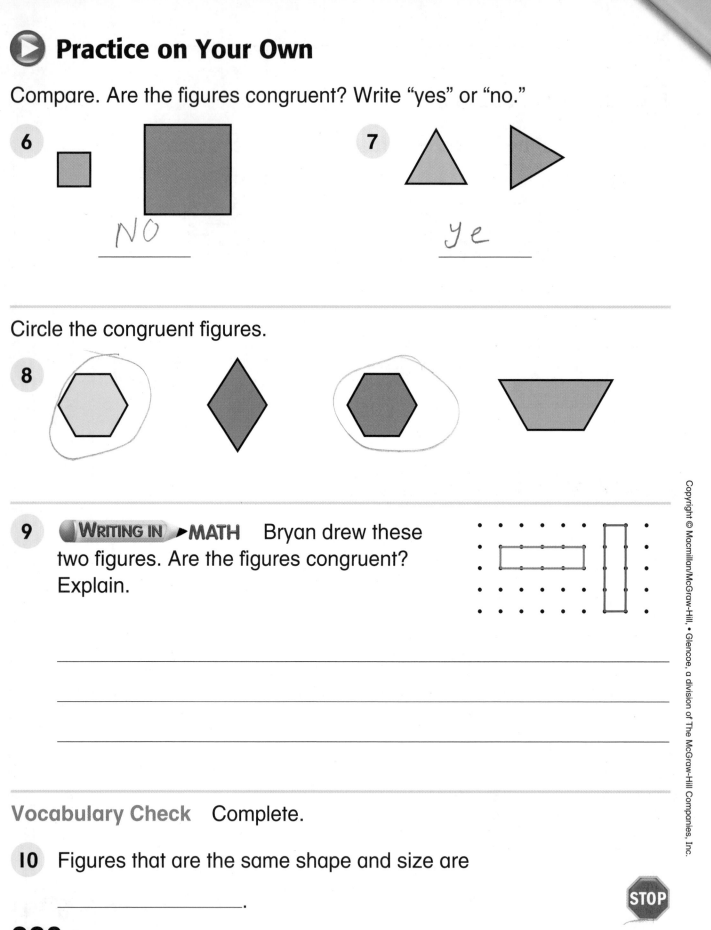

6

NO

7

ye

Circle the congruent figures.

8

9 **WRITING IN ▶MATH** Bryan drew these
two figures. Are the figures congruent?
Explain.

_____

_____

_____

Vocabulary Check Complete.

10 Figures that are the same shape and size are

_____.

STOP

Name _____

Symmetry

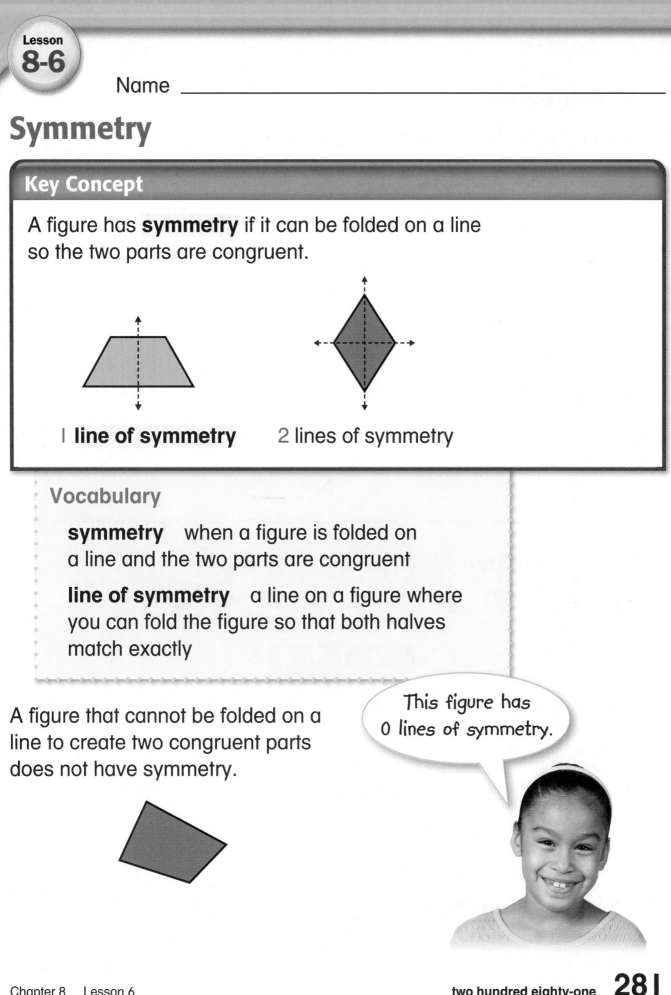

Key Concept

A figure has **symmetry** if it can be folded on a line so the two parts are congruent.

I **line of symmetry**   2 lines of symmetry

Vocabulary

symmetry   when a figure is folded on a line and the two parts are congruent

line of symmetry   a line on a figure where you can fold the figure so that both halves match exactly

A figure that cannot be folded on a line to create two congruent parts does not have symmetry.

This figure has 0 lines of symmetry.

Example

How many lines of symmetry does the heart have?

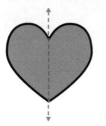

Step 1   Look at the heart.
Can you make two congruent parts? yes

Step 2   Draw the line or lines.

Step 3   Count the line(s) of symmetry. I

Answer   The heart has I line of symmetry.

Step-by-Step Practice

How many lines of symmetry does the rectangle have?

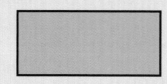

Step 1   Look at the rectangle.
Can you make two congruent parts? _____

Step 2   Draw the line or lines.

Step 3   Count the line(s) of symmetry. _____

Answer   The rectangle has _____ lines of symmetry.

Name _____

▶ Guided Practice

Count the lines of symmetry for each figure.

1

_____ line of symmetry

2

_____ lines of symmetry

Problem-Solving Practice

3 Luke drew this figure to show his age. How many lines of symmetry does this figure have?

Understand   Underline key words

Plan   Use a model.

Solve   Trace the figure.

Fold the figure to find the lines of symmetry. Draw each line of symmetry

The figure has

_____ lines of symmetry.

Check   Do the lines form congruent parts?

GO on

▶ Practice on Your Own

Count the lines of symmetry for each figure.
Draw each line of symmetry.

4

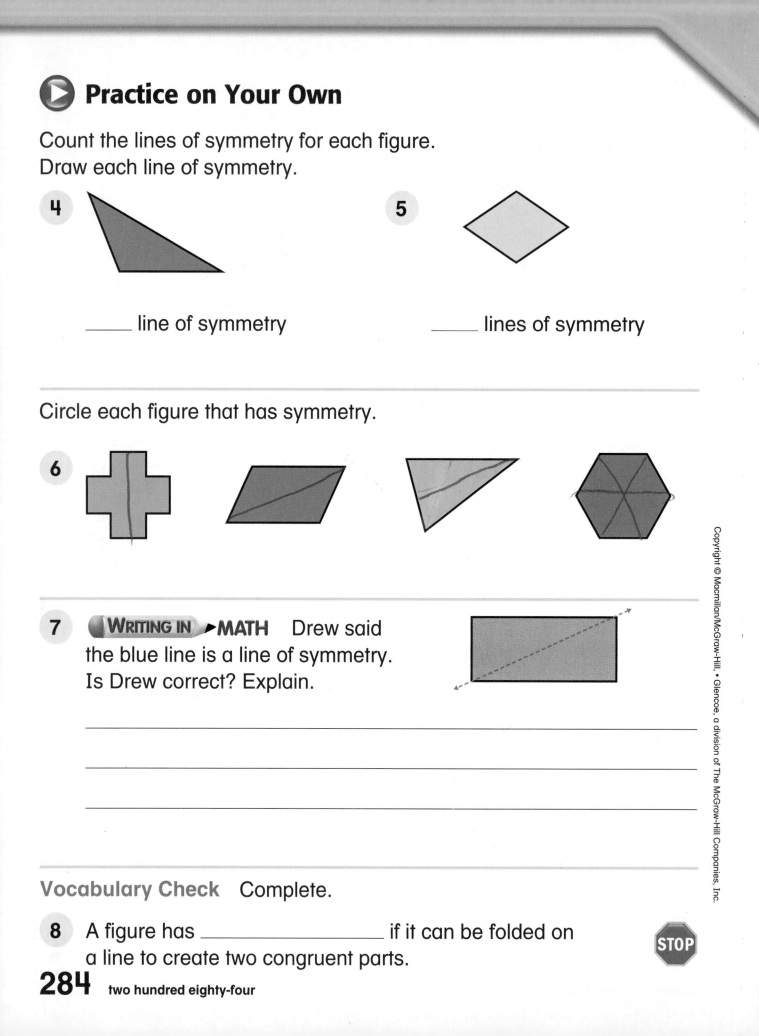

_____ line of symmetry

5

_____ lines of symmetry

Circle each figure that has symmetry.

6

---

7 **WRITING IN** ▶**MATH**   Drew said
the blue line is a line of symmetry.
Is Drew correct? Explain.

_____

_____

_____

Vocabulary Check   Complete.

8 A figure has _____ if it can be folded on
a line to create two congruent parts.

STOP

Name _____

Progress Check 3 (Lessons 8-5 and 8-6)

Compare. Are the figures congruent? Write "yes" or "no."

1 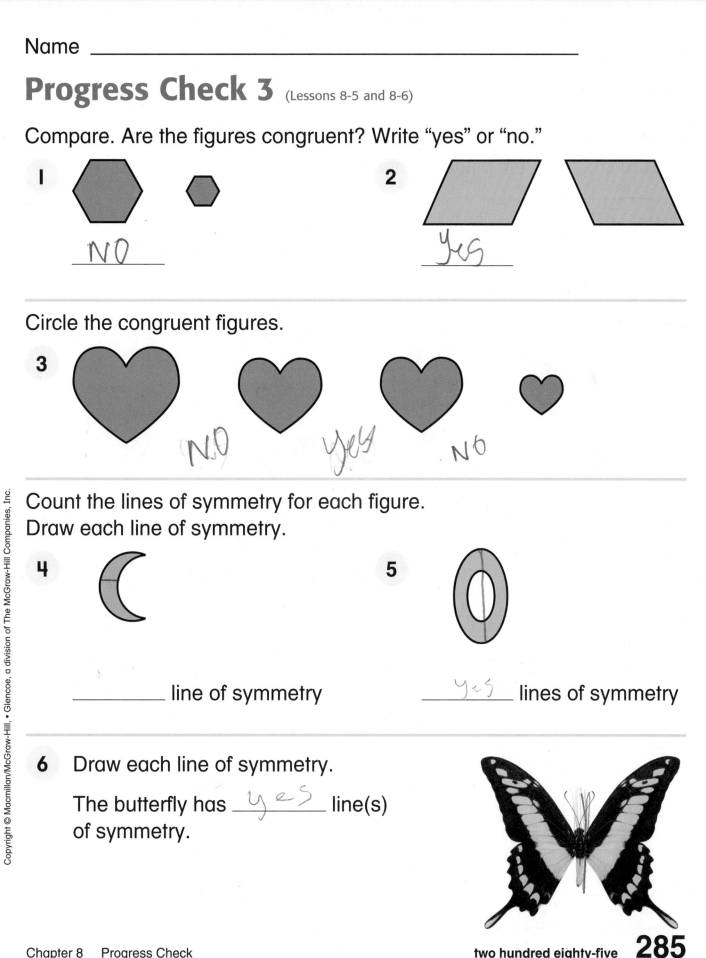 NO

2 yes

Circle the congruent figures.

3 NO   yes   NO

Count the lines of symmetry for each figure.
Draw each line of symmetry.

4 _____ line of symmetry

5 __yes__ lines of symmetry

6 Draw each line of symmetry.

The butterfly has __yes__ line(s) of symmetry.

Name _____

Draw a line to match congruent figures.
Draw a line to match a figure's lines of symmetry.

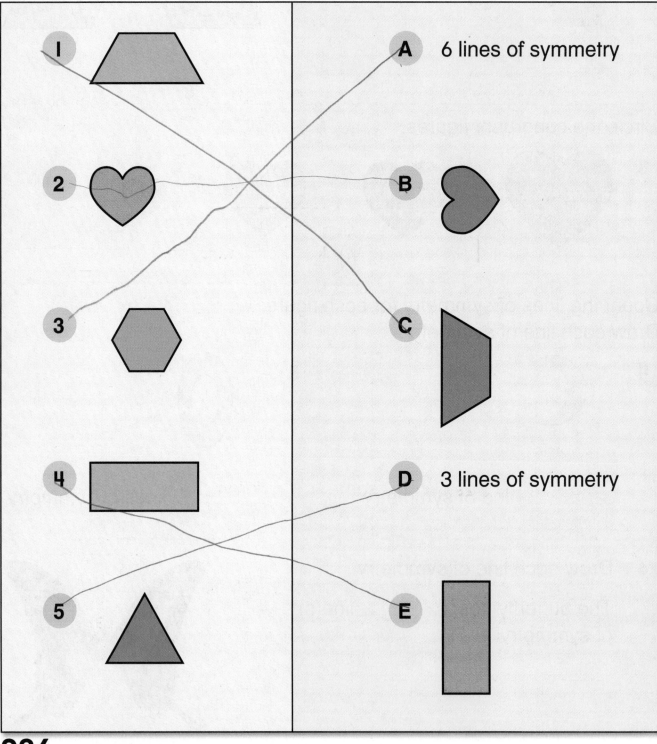

Name _____

Review

Vocabulary

Word Bank

congruent

hexagon

line of symmetry

side

Use the Word Bank to complete.

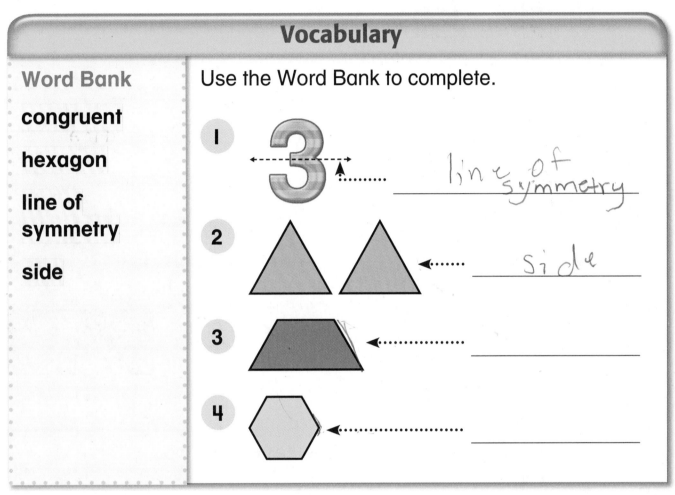

1. line of symmetry

2. side

3. _____

4. _____

▶ Concepts

5. Compare the size. Write "same" or "different."

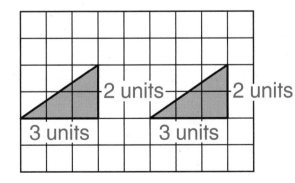

2 units 2 units
3 units 3 units

The triangles are the

_____ size.

6 Compare the shape. Write "same" or "different."

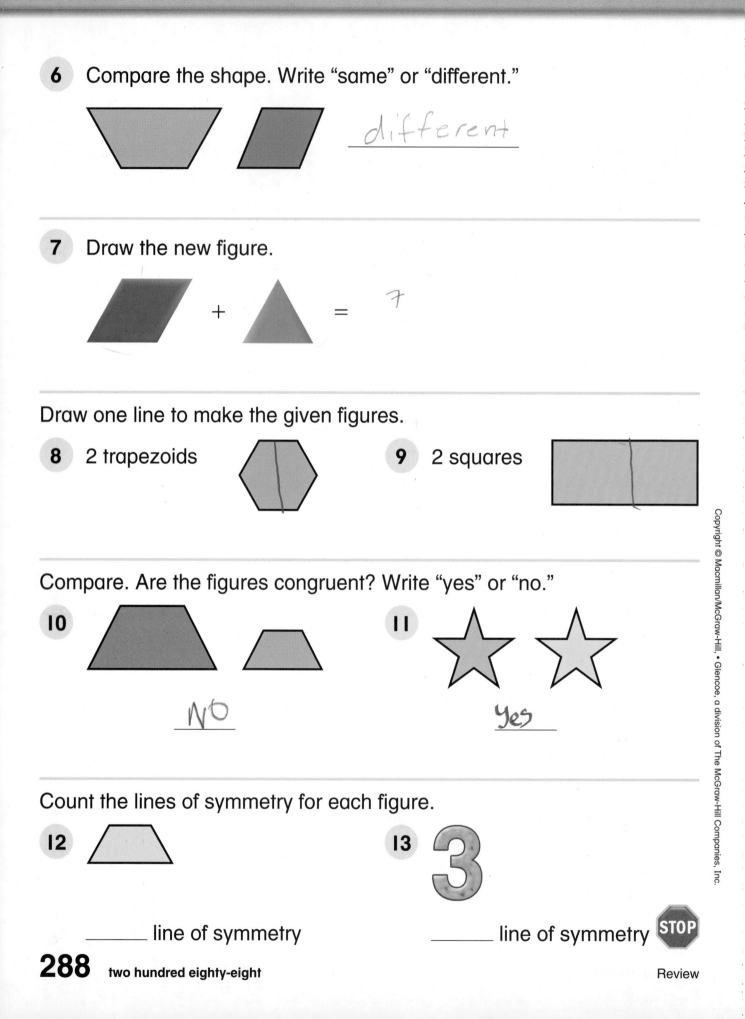

different

7 Draw the new figure.

+ = 7

Draw one line to make the given figures.

8 2 trapezoids

9 2 squares

Compare. Are the figures congruent? Write "yes" or "no."

10

NO

11

yes

Count the lines of symmetry for each figure.

12

13

_____ line of symmetry

_____ line of symmetry

STOP

Review

Name _____

Chapter Test

1 Compare the shape and size. Write "same" or "different."
Are the figures congruent?

_____Same_____ shape

_____ size

The figures are _____.

2 Circle the pattern block that completes the new figure.

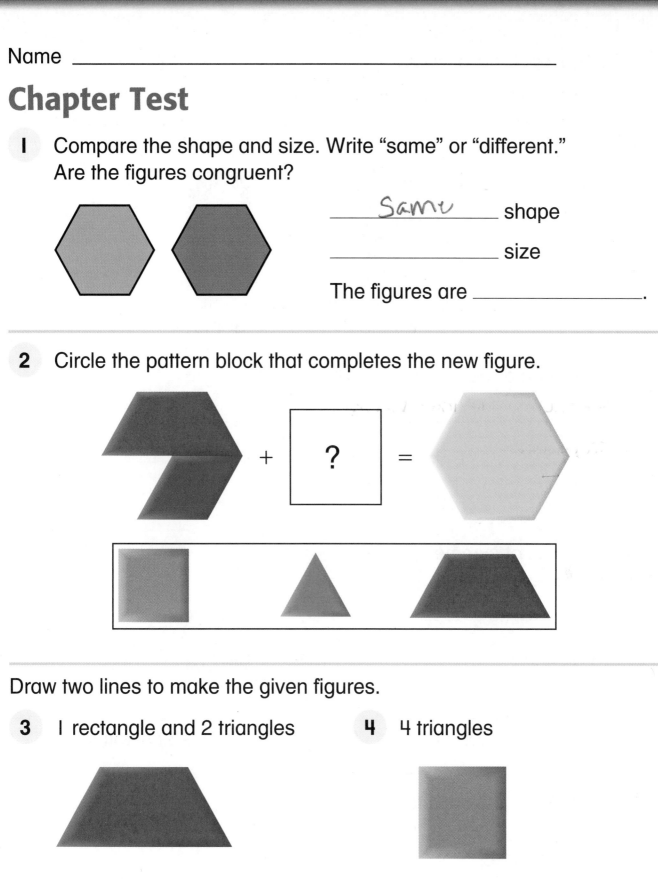

Draw two lines to make the given figures.

3 I rectangle and 2 triangles

4 4 triangles

GO on

5 Who is Correct?

Kirstie and Sarita count how many lines of symmetry a rectangle has.

A rectangle has 1 line of symmetry.

Kirstie

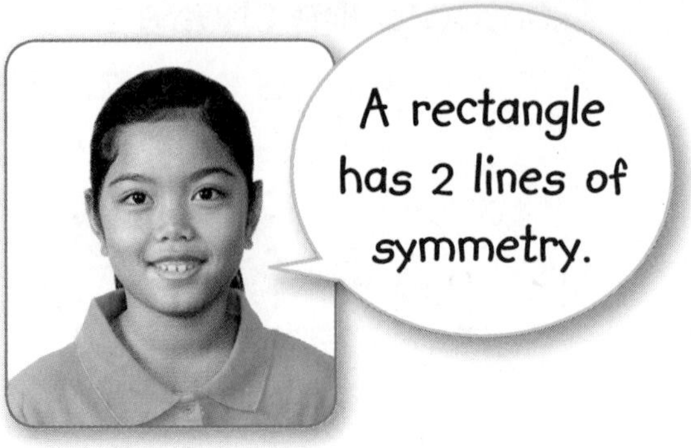

A rectangle has 2 lines of symmetry.

Sarita

Circle the correct answer. Explain.

Sarita A rectangle has 2 lines of Symmetry.

Count the lines of symmetry for each figure.
Draw each line of symmetry.

6

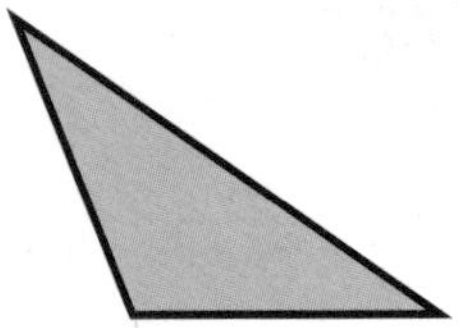

_3_ line of symmetry

7

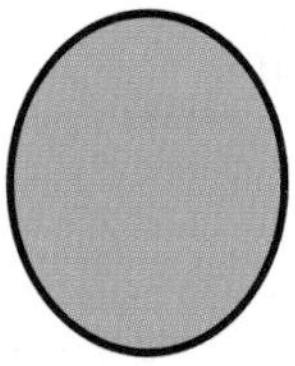

_____ lines of symmetry

---

8 A rectangle is 3 units tall and 4 units wide.
Draw a congruent rectangle.

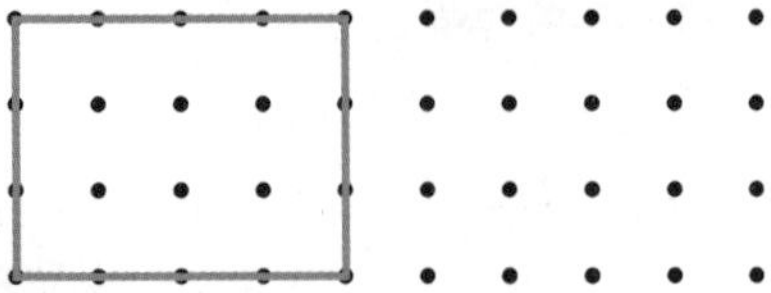

STOP

Name _____

Test Practice

Choose the correct answer.

1 Bryan drew the figures below of the following. Describe the figures.

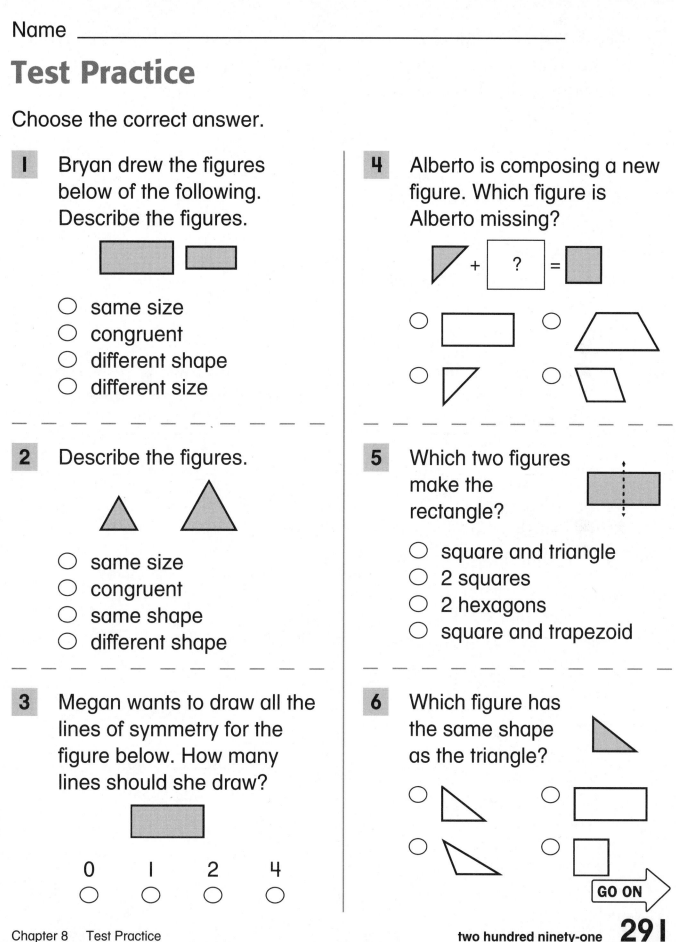

- ○ same size
- ○ congruent
- ○ different shape
- ○ different size

2 Describe the figures.

- ○ same size
- ○ congruent
- ○ same shape
- ○ different shape

3 Megan wants to draw all the lines of symmetry for the figure below. How many lines should she draw?

0	1	2	4
○	○	○	○

4 Alberto is composing a new figure. Which figure is Alberto missing?

- ○
- ○
- ○
- ○

5 Which two figures make the rectangle?

- ○ square and triangle
- ○ 2 squares
- ○ 2 hexagons
- ○ square and trapezoid

6 Which figure has the same shape as the triangle?

- ○
- ○
- ○
- ○

GO ON

7 Aidan used two figures to make the hexagon. Which two figures did Aidan use?

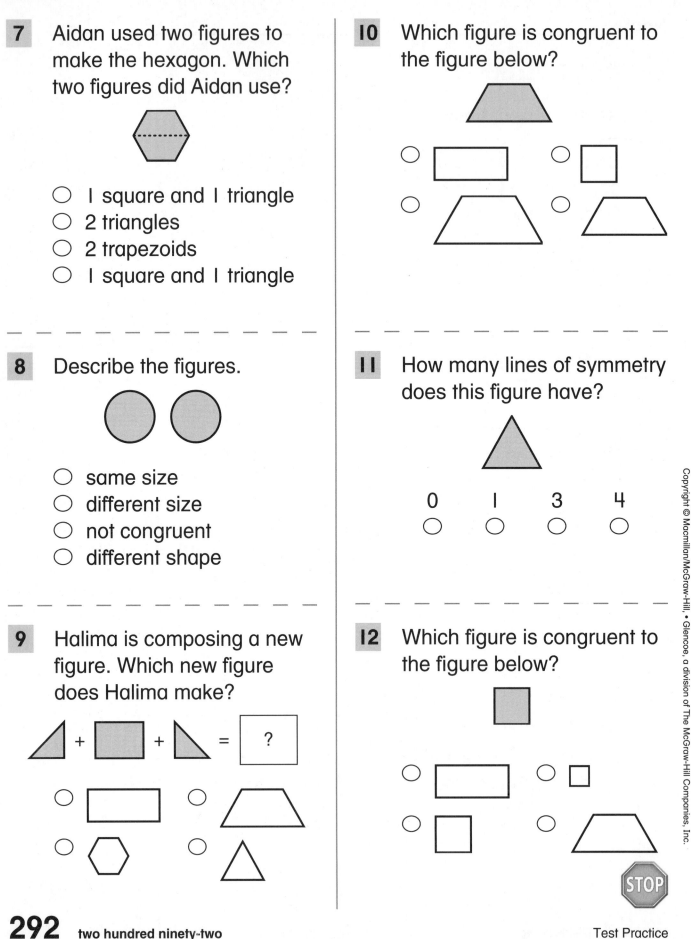

- ○ I square and I triangle
- ○ 2 triangles
- ○ 2 trapezoids
- ○ I square and I triangle

8 Describe the figures.

- ○ same size
- ○ different size
- ○ not congruent
- ○ different shape

9 Halima is composing a new figure. Which new figure does Halima make?

△ + ☐ + ◺ = ☐ ?

- ○ (rectangle)
- ○ (trapezoid)
- ○ (hexagon)
- ○ (triangle)

10 Which figure is congruent to the figure below?

- ○ (rectangle)
- ○ (square)
- ○ (trapezoid)
- ○ (trapezoid)

11 How many lines of symmetry does this figure have?

0 ○   I ○   3 ○   4 ○

12 Which figure is congruent to the figure below?

- ○ (rectangle)
- ○ (small square)
- ○ (square)
- ○ (trapezoid)

STOP

Test Practice

Home Connection

English

Spanish

Dear Family,
Today our class started
Chapter 9, Data Analysis.
In this chapter, I will learn how to read and display data on tables and graphs.

Love, _____

Estimada familia:
Hoy en clase comenzamos el
Capítulo 9, Análisis de datos.
En este capítulo aprenderé cómo leer y mostrar datos mediante tablas y gráficas.

Cariños, _____

Help at Home

You can help your child learn about data analysis at home. Ask him or her to find examples of graphs in newspapers, magazines, and textbooks. Help your child make a pictograph. Record the number of minutes he or she reads each day for one week. Create a pictograph to display the data.

Ayude en casa

Usted puede ayudar a su hijo(a) a aprender sobre el análisis de datos en casa. Pídale que encuentre ejemplos de gráficas en periódicos, revistas y libros de texto. Ayúdelo(a) a hacer una pictografía. Registre el número de minutos que él(ella) lee cada día durante una semana. Creen una pictografía que muestre los datos.

Math Online Take the chapter Get Ready quiz at macmillanmh.com.

Name _____

Get Ready

Circle the greatest number.

1   3   9   5   2   31   29   (15)

Circle the least number.

3   7   8   (4)   4   21   38   14

Find each sum or difference.

5   $6 - 4 = \underline{2}$ 6 $5 + 3 = \underline{8}$

7   $2 + 7 = \underline{9}$ 8 $8 - 1 = \underline{7}$

Find the missing numbers.

9   1   [2]   3   4   5   6   [7]   8   9   10

10   [1]   2   3   4   5   6   7   8   [9]   10

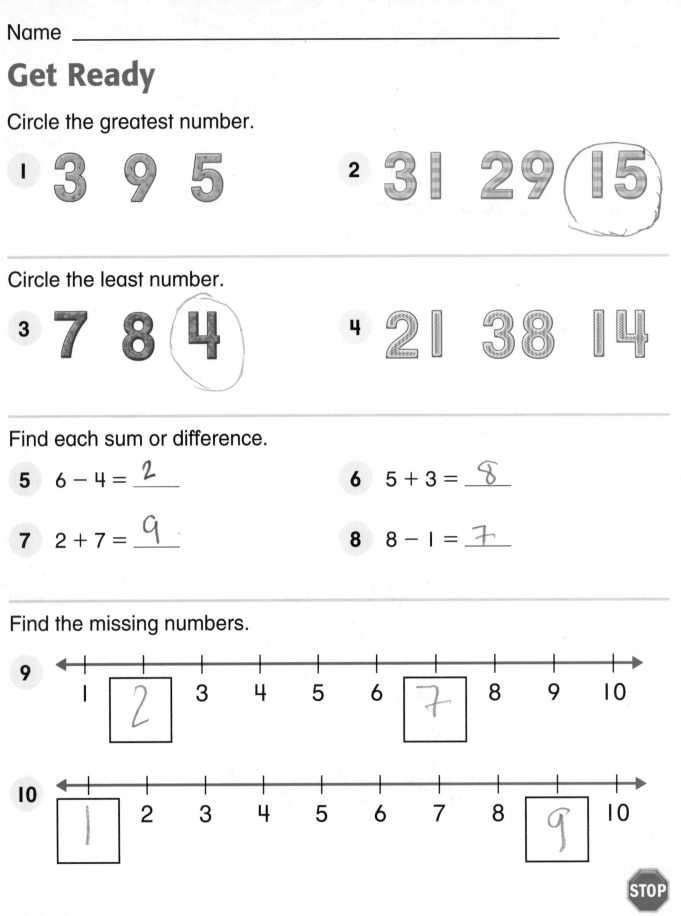

STOP

Name _____

Sort and Classify

Key Concept

You can **classify** and **sort** objects using a **Venn diagram**.

Figures can be classified by shape and color.

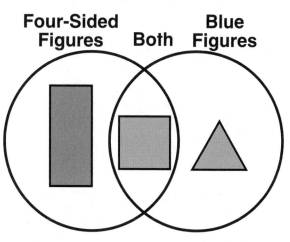

Four-Sided Figures   **Both**   **Blue Figures**

The figure on the left is a four-sided figure.
The figure on the right is a blue figure.
The figure in the middle is both a
four-sided figure and a blue figure.

Vocabulary

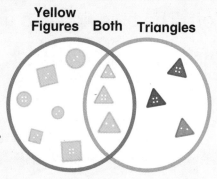

Yellow Figures   **Both**   **Triangles**

classify   to label a group of items based on a common attribute

sort   to group together like items

Venn diagram   a diagram that uses circles to organize and display data

Example

Classify and sort. Draw the figures in the Venn diagram.

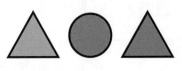

Step 1   Classify. There are triangles and red figures.

Step 2   Sort. The blue figure is a triangle.
The red circle is a red figure.
The red triangle is both.

Answer

Triangles Both Red Figures

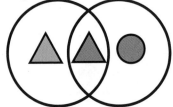

Step-by-Step Practice

Classify and sort. Draw the figures in the Venn diagram.

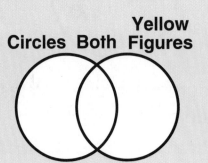

Step 1   Classify. There are _____

and _____.

Step 2   Sort. The blue figure is a _____.

The yellow square is a _____.

The yellow circle is _____.

Answer

Circles Both Yellow Figures

Name _____

▶ Guided Practice

1 Classify and sort. Draw the figures in the Venn diagram.

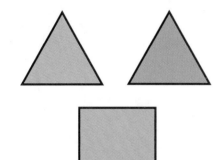

Triangles Both Green Figures

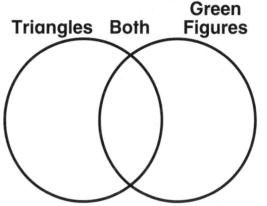

Problem-Solving Practice

2 Pamela sorted her sticker collection. How many of her stickers are stars?

Star Stickers Both Yellow Stickers

Understand	Underline key words.
Plan	Use the diagram.
Solve	Count the stickers in the **Star** section. _____
	Count the stickers in the **Both** section. _____
	Add. _____ + _____ = _____
	Pamela has _____ star stickers.
Check	Circle each star sticker in the Venn diagram. How many stars did you circle?

GO on

▶ Practice on Your Own

Classify and sort. Draw the figures in the Venn diagram.

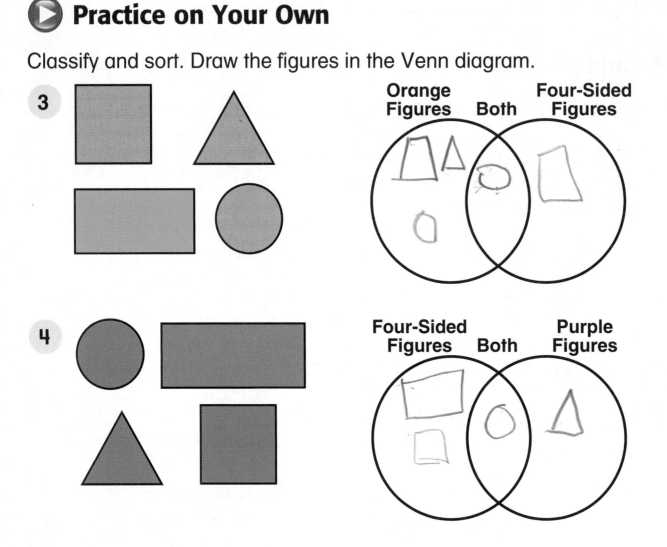

3

Orange Figures   Both   Four-Sided Figures

4

Four-Sided Figures   Both   Purple Figures

5 ◀ **WRITING IN** ▶**MATH**   Look at Exercise 4. What is another figure that could be drawn in the section with the red rectangle? Explain.

_____

_____

Vocabulary Check   Complete.

6 You can display objects that have been sorted in a

_____.

Copyright © Macmillan/McGraw-Hill • Glencoe, a division of The McGraw-Hill Companies, Inc.

Name _____

Pictographs and Picture Graphs

Key Concept

Picture Graph

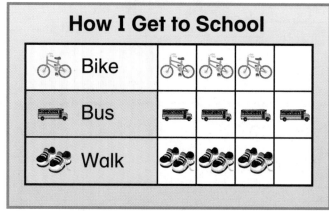

How I Get to School

🚲 Bike	🚲	🚲	🚲	
🚌 Bus	🚌	🚌	🚌	🚌
👟 Walk	👟	👟	👟	

How many students walk to school?

3 👟 = 3 students

3 students walk to school.

Pictograph

Favorite Subject

📘 Math	☺	☺	☺	☺	☺
📕 Reading	☺	☺			
📗 Science	☺	☺	☺		

Key: ☺ = 2 students

How many students chose math as their favorite subject? Skip count five times by 2.

2, 4, 6, 8, 10

10 students chose math as their favorite subject.

Vocabulary

picture graph   a graph that has different pictures to show the data

pictograph   a graph that uses the same picture or symbol to show the data

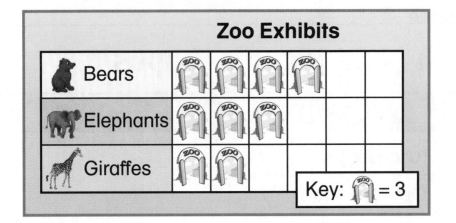

Zoo Exhibits

Key: 🏛 = 3

Example

How many bears are on exhibit at the zoo?

Step 1   Count the number of 🏛 after bears. 4

Step 2   Look at the key. 🏛 = 3

Step 3   Skip count 4 times by 3.
3, 6, 9, 12

Answer   There are 12 bears on exhibit at the zoo.

Step-by-Step Practice

How many elephants are on exhibit at the zoo?

Step 1   Count the number of 🏛 after elephants.

Step 2   Look at the key. 🏛 = 3

Step 3   Skip count 4 times by 3.
3, 6, 9

Answer   There are 9 elephants on exhibit at the zoo.

Name _____

▶ Guided Practice

Fruit Packed in Lunches on Friday

🍎 Apples	🍎	🍎	🍎	🍎				
🍌 Bananas	🍌	🍌	🍌					
🍊 Oranges	🍊	🍊						

Key: each item = 5 pieces

1 How many more bananas were packed in lunches than oranges? _____

2 How many total pieces of fruit were packed in the students' lunches on Friday? _____

Problem-Solving Practice

3 Lois made a pictograph with the key: 🍕 = 10 votes.

What does 🍕 represent?

Understand	Underline key words.
Plan	Use counters.
Solve	Separate _____ counters into _____ equal groups.

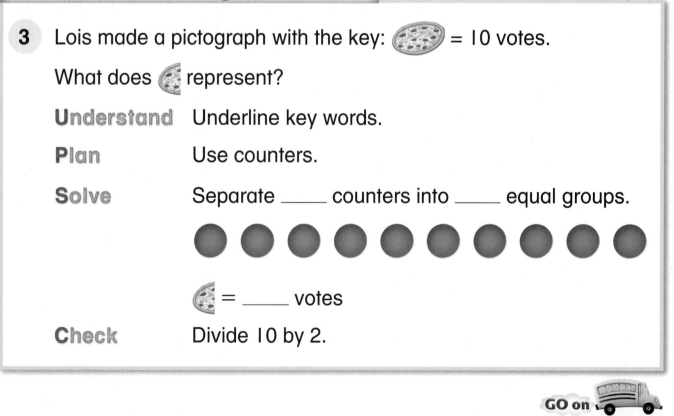

🍕 = _____ votes

Check	Divide 10 by 2.

GO on

 **Practice on Your Own**

A second grade class chose their favorite musical instruments.

Favorite Musical Instrument

🥁	Drum	☆	☆	☆				
╱	Recorder	☆	☆					
🪘	Tambourine	☆						

Key: ☆ = 10 students

4 How many students does each ☆ represent?

__2__ students

5 Which instrument is the least favorite?

__Drum__

6 How many students chose a recorder?

__2__ students

7 How many more students chose a drum than a recorder?

__>2__ students

8 🖊 **WRITING IN ►MATH** How many students are in the second grade? How do you know?

_____

_____

Vocabulary Check Complete.

9 A _____ uses the same picture or symbol to show data.

STOP

Name _____

Progress Check 1 (Lessons 9-1 and 9-2)

1 Classify and sort. Draw the figures in the Venn diagram.

____ ____ ____

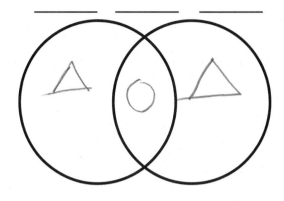

Use the graphs to answer each question.

2 How many students chose a fish as their favorite pet?

___3___ students

3 How many more students chose a dog than cat?

___4___ students

4 How many students play football?

___2___ students

5 Which sport has the least players?

___Soccer___

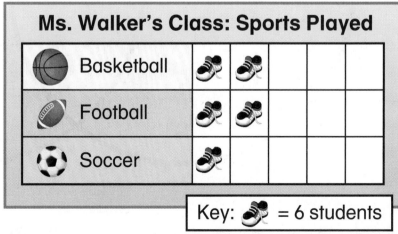

Mr. Doscher's Class: Favorite Pets

Cat				
Dog				
Fish				

Ms. Walker's Class: Sports Played

Basketball				
Football				
Soccer				

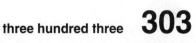

Key: 👟 = 6 students

Name _____

Work with a partner to practice classifying
and sorting objects.

Listen as your teacher instructs how to play.

Materials
pattern blocks
Venn diagram

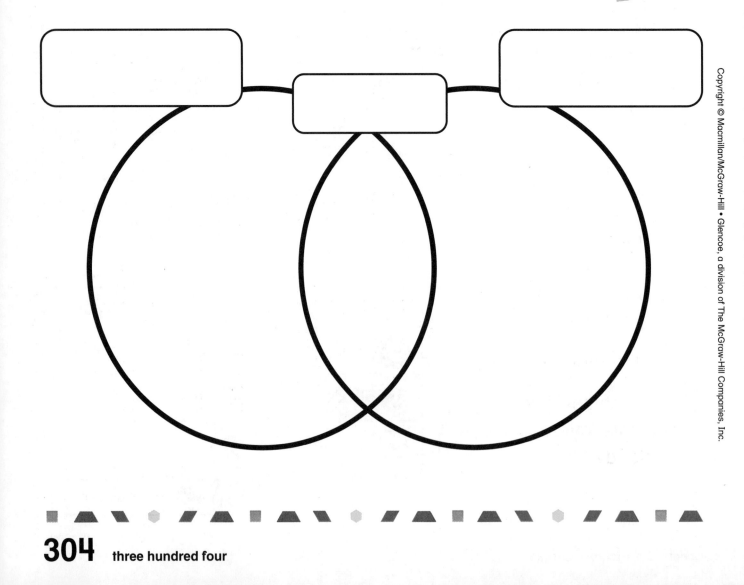

Name _____

Read Tables

Key Concept

A **table** can display **data**.

Fruit at a Farm Stand		
Fruit	Tally	Total
Apples	ЖЖ ЖЖ ‖	12
Peaches	‖‖	3
Pears	ЖЖ	5

The title of the table is Fruit at a Farm Stand.
The columns show the type of Fruit and the Total number.
There are 12 apples, 3 peaches, and 5 pears.

Vocabulary

table   a way to organize data

data   numbers or symbols that show information

tally mark   a mark used to record data collected in a survey

ЖЖ ‖
tally marks

> There are 17 apples and pears at the farm stand.

You can use a table to answer questions.

How many apples and pears are at the farm stand?

Example

How many students ate eggs for breakfast?

Breakfast Food		
Food	Tally	Number of Students
Cereal	~~HHT~~ ~~HHT~~ IIII	14
Eggs	~~HHT~~ ~~HHT~~	10
Pancakes	~~HHT~~ III	8

Step 1   Find Eggs in the Food column.

Step 2   Look across the row to the Tally column.
There are 10 tallies.

Answer   10 students ate eggs for breakfast.

Step-by-Step Practice

How many students chose pizza for lunch?

Lunch Count		
Food	Tally	Number of Students
Hamburger	~~HHT~~ ~~HHT~~ IIII	14
Pizza	~~HHT~~ ~~HHT~~ ~~HHT~~ ~~HHT~~	20
Turkey Sub	~~HHT~~ ~~HHT~~ III	13

Step 1   Find __Pizza__ in the Food column.

Step 2   Look across the row to the Tally column.

There are __20__ tallies.

Answer   __20__ students chose pizza for lunch.

Name _____

▶ Guided Practice

1 How many people chose

   blue? __6__ people

2 How many people chose

   green? __4__ people

3 How many more people
   chose red than green?

   __9__ – __4__ = __6__ more people

Favorite Color		
Color	Tally	Number of People
Blue	⊬⊬Ⅰ	6
Green	ⅠⅠⅠⅠ	4
Red	⊬⊬ⅠⅠⅠⅠ	9

Problem-Solving Practice

4 How many people chose
   spring and winter as their
   favorite season?

Favorite Season		
Season	Tally	Number of People
Fall	⊬⊬ ⊬⊬	10
Spring	⊬⊬ ⊬⊬ ⅠⅠ	12
Summer	⊬⊬ ⊬⊬ ⊬⊬	15
Winter	⊬⊬ ⅠⅠⅠ	7

Understand Underline
key words.

Plan Write a
number
sentence.

Solve _____ spring votes + _____ winter votes

_____ people chose spring and winter.

Check Count the tally marks for spring
and winter.

GO on

Practice on Your Own

Playground Equipment at Shull Park		
Equipment	Tally	Total
Seesaws	IIII	4
Slides	HHT I ~	6
Swings	HHT III	+ 8

5 How many swings are on the playground? __8__ swings

6 How many slides are on the playground? __6__ slides

7 How many more swings are on the playground
than seesaws? __4__ more   8 − 4 =

8 How many swings, seesaws, and slides are on
the playground in all? __18__ in all

9 WRITING IN ▶MATH There
are 20 students in Cameron's
class. How many boys are in
his class? Explain.

Students in Cameron's Class		
Gender	Tally	Total
Boys	?	?
Girls	HHT HHT I	11

_____

_____

Vocabulary Check Complete.

10 A _____ has a title and labels and is used to
organize data.

308 three hundred eight

Name _____

Read Bar Graphs

Key Concept

You can read **data** from a **bar graph**.

Each bar shows a different snack food.
The length of the bar shows the number
of students who voted for the snack.

Vocabulary

data   numbers or symbols
that show information

bar graph   a graph that
uses bars to show data

Bar graphs can be shown
another way. The bars on this
graph look different but show
the same data.

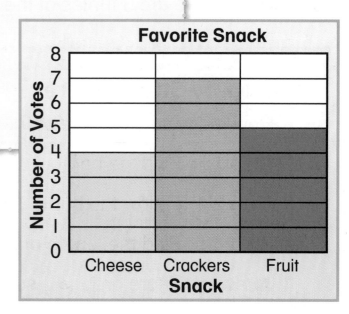

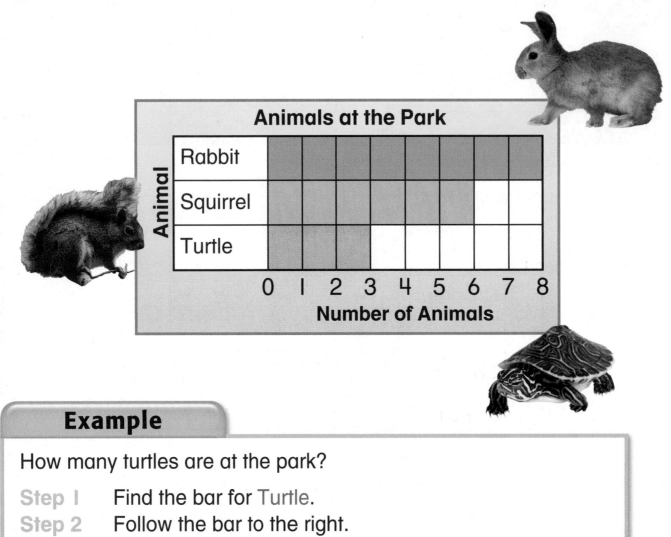

Animals at the Park

Animal	Number of Animals
Rabbit	(bar to 8)
Squirrel	(bar to 6)
Turtle	(bar to 3)

0 1 2 3 4 5 6 7 8

Number of Animals

Example

How many turtles are at the park?

Step 1   Find the bar for Turtle.

Step 2   Follow the bar to the right.

Step 3   Find the length of the bar. 3

Answer   There are 3 turtles at the park.

Step-by-Step Practice

How many squirrels are at the park?

Step 1   Find the bar for _Squirrel_.

Step 2   Follow the bar to the right.

Step 3   Find the length of the bar. _6_

Answer   There are _6_ squirrels at the park.

Name _____

▶ Guided Practice

1 How many daisies are in the garden?

___7___ daisies

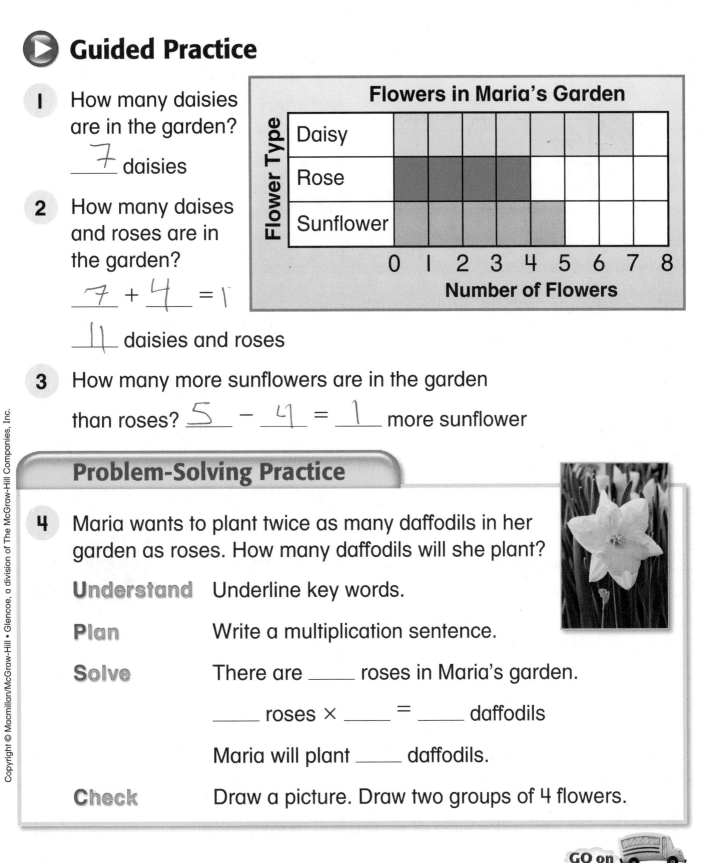

Flowers in Maria's Garden

Flower Type: Daisy, Rose, Sunflower

Number of Flowers: 0 1 2 3 4 5 6 7 8

2 How many daises and roses are in the garden?

___7___ + ___4___ = ___1___

___11___ daisies and roses

3 How many more sunflowers are in the garden than roses? ___5___ – ___4___ = ___1___ more sunflower

Problem-Solving Practice

4 Maria wants to plant twice as many daffodils in her garden as roses. How many daffodils will she plant?

Understand Underline key words.

Plan Write a multiplication sentence.

Solve There are _____ roses in Maria's garden.

_____ roses × _____ = _____ daffodils

Maria will plant _____ daffodils.

Check Draw a picture. Draw two groups of 4 flowers.

GO on

Practice on Your Own

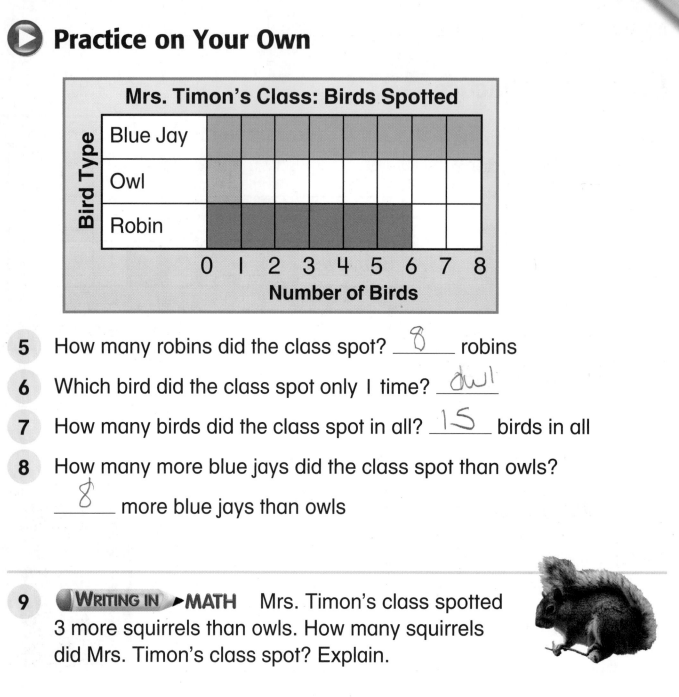

Mrs. Timon's Class: Birds Spotted

Bird Type	Number of Birds
Blue Jay	8
Owl	1
Robin	6

(scale: 0 1 2 3 4 5 6 7 8)

5   How many robins did the class spot? ___8___ robins

6   Which bird did the class spot only 1 time? ___owl___

7   How many birds did the class spot in all? ___15___ birds in all

8   How many more blue jays did the class spot than owls?

___8___ more blue jays than owls

9   **WRITING IN ▸MATH**   Mrs. Timon's class spotted 3 more squirrels than owls. How many squirrels did Mrs. Timon's class spot? Explain.

_____

_____

Vocabulary Check   Complete.

10   A graph that uses bars to show data is called a

_____.

STOP

Name _____

Progress Check 2 (Lessons 9-3 and 9-4)

Use the table to answer each question.

1 How many water bottles were recycled?

____17____ water bottles

2 What item had 9 recycled?

Newspapers

Recycled Items		
Item	Tally	Number of Items
Cans	~~HHt~~ ~~HHt~~ II	12
Newspapers	~~HHt~~ IIII	9
Water Bottles	~~HHt~~ ~~HHt~~ ~~HHt~~ II	17

+ 39

3 How many items were recycled in all? ___39___ items in all

Use the bar graph to answer each question.

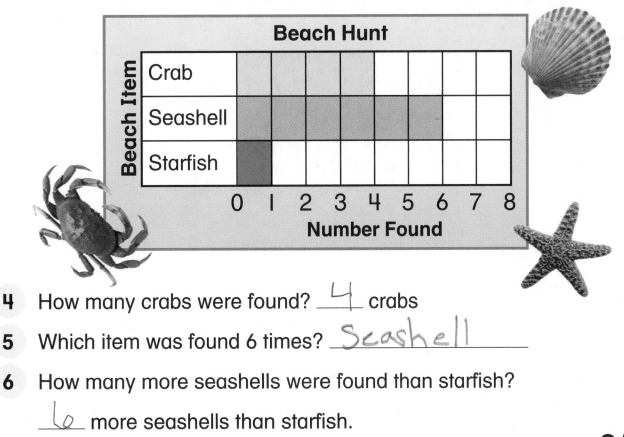

4 How many crabs were found? __4__ crabs

5 Which item was found 6 times? _Seashell_

6 How many more seashells were found than starfish?

__6__ more seashells than starfish.

Name _____

Read the ZE-BAR Graph!

Use the bar graph to complete each sentence.

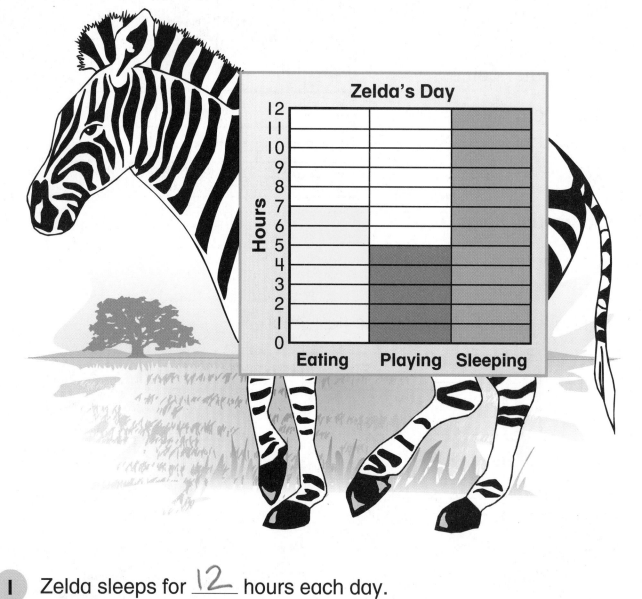

Zelda's Day

1 Zelda sleeps for __12__ hours each day.

2 Zelda spends the least time __Playing__.

3 Zelda sleeps __12__ hours more than she plays.

4 Zelda spends __19__ hours eating and sleeping each day.

5 Zelda spends 2 more hours eating than __Playing__.

314 three hundred fourteen

Name _____

Make Bar Graphs

Key Concept

You can use **data** from a **table** to make a **bar graph**.

Weekend Activity

Activity	Tally	Number of People
Movie	‖‖‖ ‖‖‖	9
Museum	‖‖‖	3
Park	‖‖‖ ‖‖	7

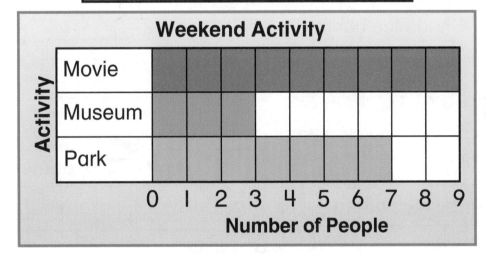

Weekend Activity

The length of each bar matches the numbers in the table.

Vocabulary

data   numbers or symbols that show information

table   a way to organize data

bar graph   a graph that uses bars to show data

Example

Use the table to complete the bar graph.

Dogs Washed		
Dog Type	Tally	Number of Dogs
Collie	IIII	4
Lab	HHT	5
Poodle	II	2

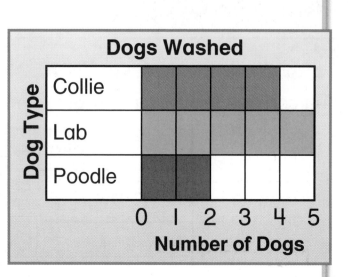

Dogs Washed

Step 1   Shade the bar for Collie to show 4.
Step 2   Shade the bar for Lab to show 5.
Step 3   Shade the bar for Poodle to show 2.

Step-by-Step Practice

Use the table to complete the bar graph.

Animals Groomed		
Animal Type	Tally	Number of Animals
Cat	III	3
Dog	HHT	5
Horse	I	1

Animals Groomed

Step 1   Shade the bar for Cat to show _3_.

Step 2   Shade the bar for Dog to show _5_.

Step 3   Shade the bar for Horse to show _1_.

Name _____

 Guided Practice

1 Use the table to complete the bar graph.

Favorite Fruit		
Fruit	Tally	Number of People
Banana	|||	3
Orange	ЖЖ		6
Pear	ЖЖ ||	7

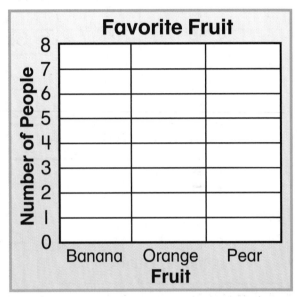

Favorite Fruit

Number of People: 8 7 6 5 4 3 2 1 0

Banana   Orange   Pear

Fruit

Problem-Solving Practice

2 How many more people chose pears than bananas?

Understand Underline key words.

Plan Use the bar graph.

Solve How many people chose pears? _____

How many people chose bananas? _____

_____ – _____ = _____ more people

Check Work backward. How many spaces taller is "Pear" than "Banana"?

GO on 🚌

▶ Practice on Your Own

Use the tables to complete each bar graph.

3

Siblings		
Age	Tally	Number of Siblings
Older	|||	3
Twin			1
Younger	||||	4

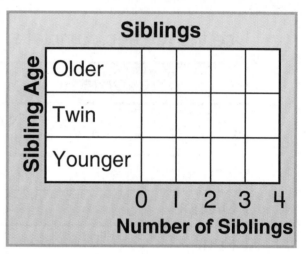

4

Hair Color		
Color	Tally	Number of People
Blonde	⊬⊬	5
Brown	⊬⊬ ||	7
Red	||	2

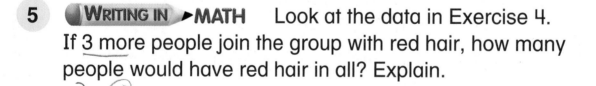

5 **WRITING IN ▶MATH** Look at the data in Exercise 4. If 3 more people join the group with red hair, how many people would have red hair in all? Explain.

2 + 3 = 5

Vocabulary Check   Complete.

6 You can display _____ from a table in a bar graph.

318 three hundred eighteen

Name _____

Line Plots

Key Concept

A **line plot** is another way to show **data**.

The line plot below shows the number of homeruns kicked by kickball players.

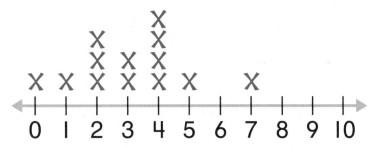

Homeruns Scored

Each X equals 1 kickball player.

There are 2 Xs above the 3.
So, 2 kickball players kicked 3 homeruns.

Vocabulary

line plot   a graph that uses columns of Xs above a number line to show frequency of data

data   numbers or symbols that show information

There are 0 Xs above the 6.
So, 0 kickball players kicked 6 homeruns.

Example

This line plot shows the number of points scored on a game show. How many people scored 3 points?

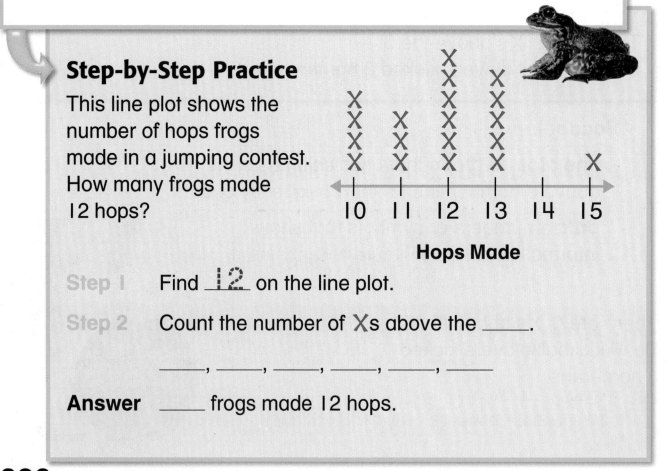

Points Scored

Step 1   Find 3 on the line plot.

Step 2   Count the number of Xs above the 3.
1, 2, 3, 4, 5

Answer   5 people scored 3 points.

Step-by-Step Practice

This line plot shows the number of hops frogs made in a jumping contest. How many frogs made 12 hops?

Hops Made

Step 1   Find __12__ on the line plot.

Step 2   Count the number of Xs above the _____.

_____, _____, _____, _____, _____, _____

Answer   _____ frogs made 12 hops.

Name _____

▶ Guided Practice

This line plot shows the
number of baskets made
by players during a
game. Use the line plot to
answer each question.

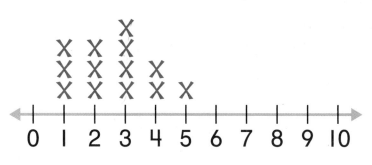

Baskets Made

1 How many players made 4 baskets? __2__ players

2 How many more players made 3 baskets than 5 baskets?

__5__ more players

3 How many baskets did the greatest number of players make?

__4__ baskets

Problem-Solving Practice

4 How many players made baskets in all?

Understand   Underline key words.

Plan   Write a number sentence.

Solve   Find the sum of the number of players
that made 1, 2, 3, 4, and 5 baskets.

$3 + 3 + 4 + 2 + 1 = 13$ players

Check   Count the total number of Xs on the line plot.

GO on

▶ Practice on Your Own

This line plot shows the number of goals made by hockey players over a season. Use the line plot to answer each question.

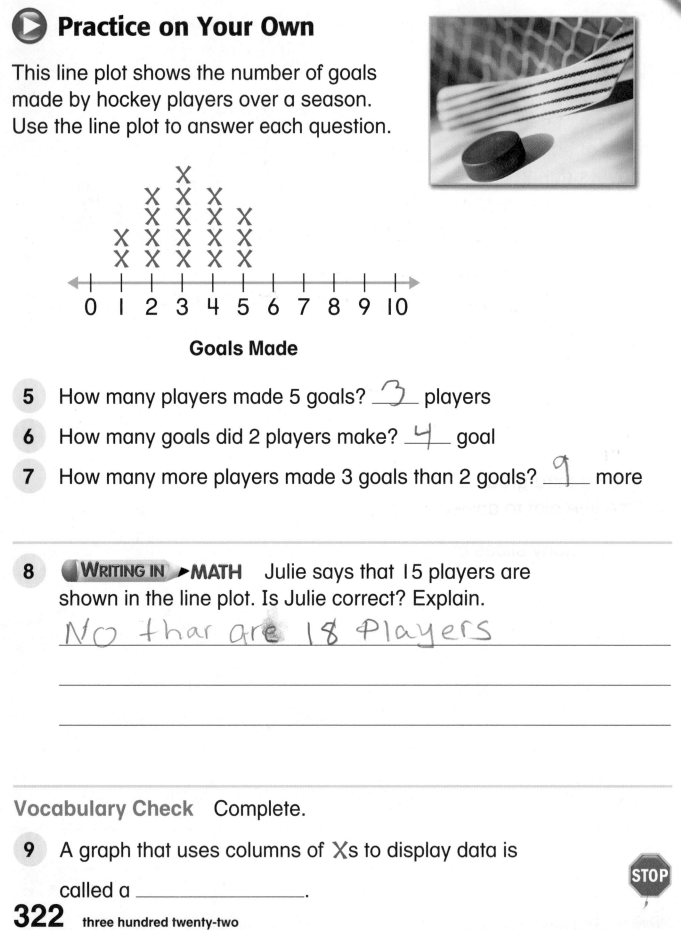

Goals Made

5 How many players made 5 goals? __3__ players

6 How many goals did 2 players make? __4__ goal

7 How many more players made 3 goals than 2 goals? __9__ more

8 **WRITING IN ▶MATH**   Julie says that 15 players are shown in the line plot. Is Julie correct? Explain.

No thar are 18 Players

_____

_____

Vocabulary Check   Complete.

9 A graph that uses columns of Xs to display data is

called a _____.

Name _____

Progress Check 3 (Lessons 9-5 and 9-6)

I Use the data in the table to complete the bar graph.

Favorite Pizza Topping		
Topping	Tally	Number of Votes
Cheese	⟍⟍⟍⟍ l	6
Pepperoni	lll	3
Sausage	⟍⟍⟍⟍	5

Favorite Pizza Topping

Topping

Cheese

Pepperoni

Sausage

0 1 2 3 4 5 6
Number of Votes

This line plot shows the number of pizza slices eaten by a third grade class for lunch one day. Use the line plot to answer each question.

2 How many slices of pizza did most students eat?

___2___ slices

3 How many students ate 3 slices of pizza?

___4___ students

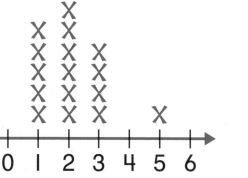

Slices of Pizza Eaten

4 There are no Xs over the 4. What does this mean?

NO 4

Name _____

Make a Lion-Plot

Use the clues to place the Xs on the line plot.
The line plot shows the lengths of 9 lion tails.

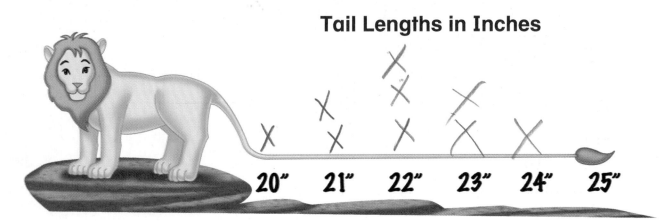

Tail Lengths in Inches

20"   21"   22"   23"   24"   25"

1 Two lions have a 21-inch long tail.

2 There is 1 lion with a tail 20 inches long.

3 Three lions have a 22-inch long tail.

4 No lion has a tail 25 inches long.

5 There is one lion with a tail 24 inches long.

6 Two lions have a 23-inch long tail.

Review

Vocabulary

Word Bank

bar graph

line plot

picture graph

Use the Word Bank to complete.

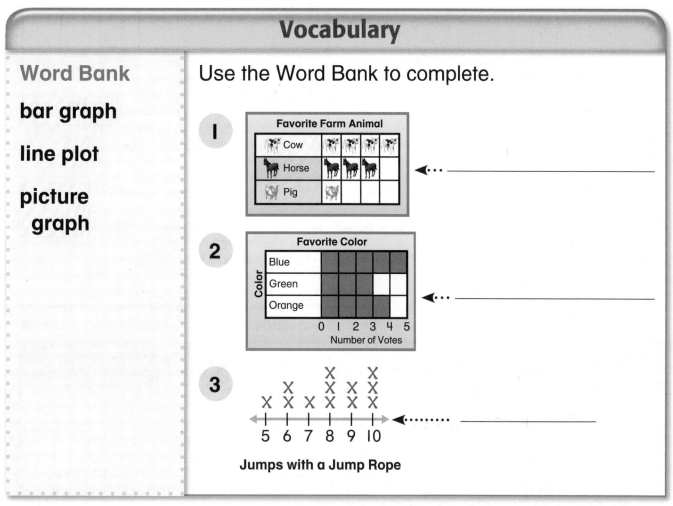

1

Favorite Farm Animal

Cow					
Horse					
Pig					

←··· _____

2

Favorite Color

Color

Blue					
Green					
Orange					

0 1 2 3 4 5
Number of Votes

←··· _____

3

```
                X
        X   X   X   X
    X   X   X   X   X
    |   |   |   |   |   |
    5   6   7   8   9   10
```

Jumps with a Jump Rope

←········ _____

▶ Concepts

4 How many veggie sandwiches were sold?

_____

Number of Sandwiches Sold		
Sandwich	Tally	Number Sold
Ham	卌 卌 ||	12
Tuna	卌 ||||	9
Veggie	卌 卌 卌 ||	17

GO on

5 Which dog biscuit flavor sold the greatest number? __5__

6 How many peanut butter biscuits were sold? __3__

Number of Dog Biscuits Sold

Milk	🦴	🦴	🦴	🦴
Mint	🦴	🦴		
Peanut Butter	🦴	🦴	🦴	

Key: 🦴 = 5 biscuits

---

7 Use the table to complete the bar graph and answer each question.

Favorite Ice Cream Flavor

Flavors	Tally	Number of Votes
Chocolate	IIII I	6
Strawberry	IIII	5
Vanilla	IIII	4

Favorite Ice Cream Flavor

8 How many people chose vanilla? __4__ people

9 How many more people chose chocolate than strawberry?

__1☐__ more

10 How many people were surveyed in all?

__15__ people in all

Review

Name _____

Chapter Test

1   Classify and sort. Draw the figures in the Venn Diagram.

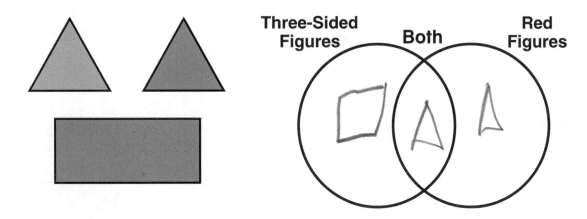

Use the bar graph to answer each question.

Favorite Smoothie Flavor

2   How many people chose banana? __2__

3   How many more people chose raspberry than banana? __10__

4   How many people chose coconut and banana? __5__

5   How many people voted in all? __13__

GO on

6 Who is Correct?

Ethan and Matt say how many goldfish are in the tank.

> There are 4 goldfish in the tank.

Ethan

Fish in a Tank

Goldfish	🐟	🐟	🐟	🐟		
Rockfish	🐟	🐟	🐟	🐟	🐟	🐟

Key: 🐟 = 2 fish

> There are 8 goldfish in a tank.

Matt

Circle the correct answer. Explain.

Ethan

_____

_____

_____

This line plot shows the number of fish caught by local fishers. Use the line plot to answer each question.

7 How many fish did the greatest number of fishers catch?

_____ fish

8 How many fishers caught

30 fish? _20_ fishers

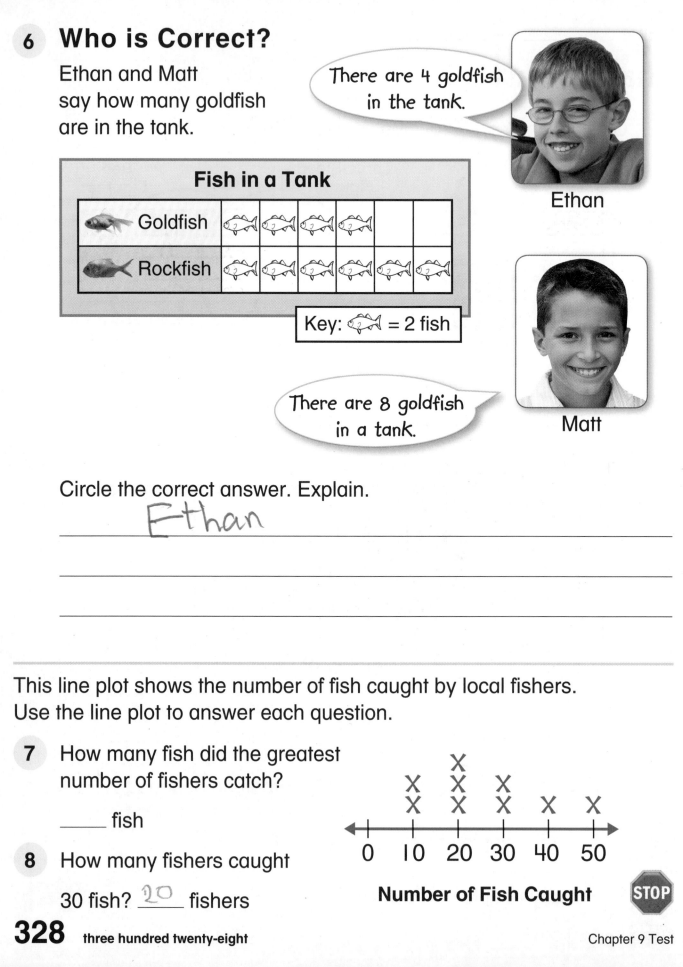

Number of Fish Caught

STOP

328 three hundred twenty-eight

Copyright © Macmillan/McGraw-Hill • Glencoe, a division of The McGraw-Hill Companies, Inc.

Name _____

Test Practice

Choose the correct answer.

1 How many lion cookie cutters are there?

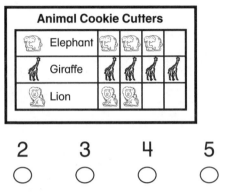

| 2 | 3 | 4 | 5 |
| ○ | ○ | ○ | ○ |

2 Look at the picture graph in Question 1. How many more giraffe than elephant cookie cutters are there?

| 1 | 2 | 3 | 4 |
| ○ | ○ | ○ | ○ |

3 How many people own 5 DVDs?

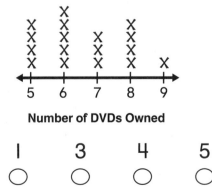

| 1 | 3 | 4 | 5 |
| ○ | ○ | ○ | ○ |

4 How many water bottles were recycled?

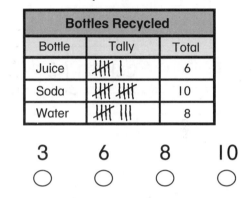

| 3 | 6 | 8 | 10 |
| ○ | ○ | ○ | ○ |

5 Look at the table in Question 4. How many juice and soda bottles were recycled?

| 6 | 10 | 14 | 16 |
| ○ | ○ | ○ | ○ |

6 How many more people chose picnics than hiking?

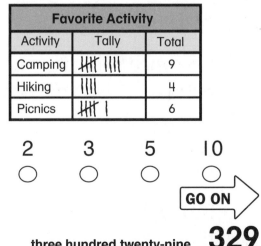

| 2 | 3 | 5 | 10 |
| ○ | ○ | ○ | ○ |

GO ON

7 Look at the table in Question 6. Which activity had the greatest number of votes?

○ Picnics   ○ Camping
○ Hiking   ○ Fishing

8 How many foxes are in the forest?

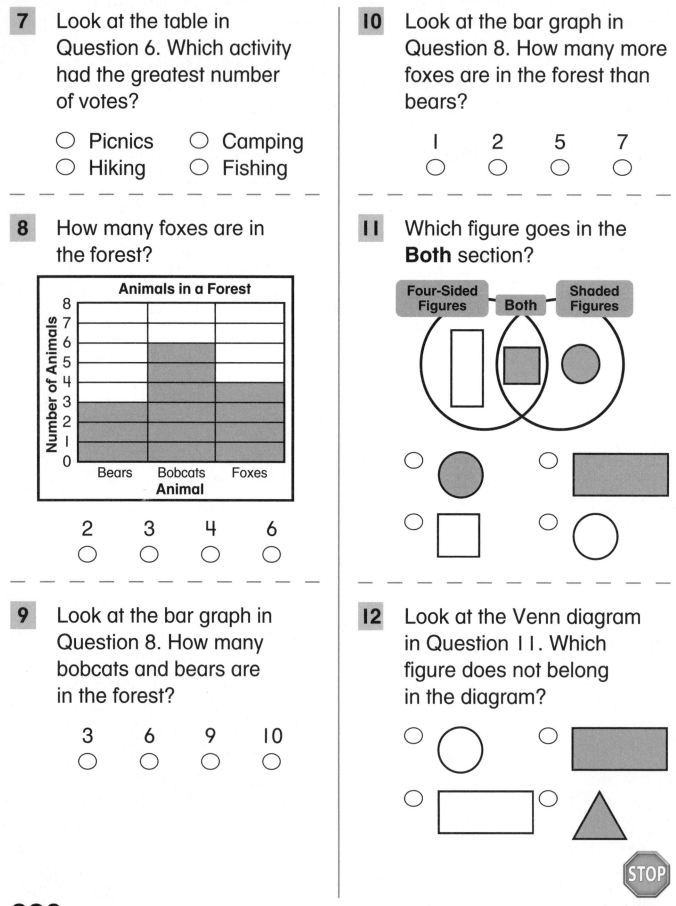

Animals in a Forest

Number of Animals

Bears   Bobcats   Foxes
Animal

    2   3   4   6
    ○   ○   ○   ○

9 Look at the bar graph in Question 8. How many bobcats and bears are in the forest?

    3   6   9   10
    ○   ○   ○   ○

10 Look at the bar graph in Question 8. How many more foxes are in the forest than bears?

    1   2   5   7
    ○   ○   ○   ○

11 Which figure goes in the **Both** section?

Four-Sided Figures   Both   Shaded Figures

12 Look at the Venn diagram in Question 11. Which figure does not belong in the diagram?

STOP

Test Practice